WINESENSE

The Three Keys to Understanding Wine

Bob Desautels

FriesenPress

Suite 300 - 990 Fort St
Victoria, BC, Canada, V8V 3K2
www.friesenpress.com

Copyright © 2015 by Bob Desautels
First Edition — 2015

Illustrations by Gillian Wilson.

ISBN
978-1-4602-5828-6 (Hardcover)
978-1-4602-5829-3 (Paperback)
978-1-4602-5830-9 (eBook)

1. Cooking, Beverages, Wine & Spirits

Distributed to the trade by The Ingram Book Company

Table of Contents

For Susanne Kaye

I am thinking what the grapes are thinking
become part of their purple mentality
that is

I am satisfied with the sun and
eventual fermenting bubble-talk together
then transformed and glinting with coloured lights in

a GREAT JERABOAM

that booms inside from the land beyond the world

—Al Purdy, "The Winemaker's Beat-Étude"

Acknowledgments & Thanks

First and foremost, I need to thank my wife, Sue, who continually read and commented on the early handwritten drafts of the manuscripts. Her encouragement and positive remarks were central to the whole process.

The most instrumental person in making this book a reality was Louise McMullen, a long-term employee at my company. Without her help in putting the book into Word and her ongoing spot editing, I don't think this project would have ever been completed. Louise's dedication to "getting it done" inspired me to keep forging ahead. She's an amazingly diligent and organized person.

My good friend Carolyn Pletsch was almost singlehandedly responsible for the overall readability of the book. She transformed my sometimes quirky prose into a much smoother and coherent form. And like a real professional, Carolyn was always quick with her feedback and constructive in her criticisms. I was lucky to have such a terrific editor.

The book has been nicely enhanced thanks to the fine drawings of Gillian Wilson. She was very clever in translating some of my directions into pictures. Her drawings helped to show some of the main aspects of wine production and related topics. Thankfully, I also received some help on the technical aspects of winemaking. Canadian wine industry pioneer and legend Don Ziraldo was kind enough to read the manuscript while travelling in Portugal and Italy. His email posts from abroad were always welcomed and valued. I also owe a special thanks to Louise Engel, a former student of mine, and David Johnson — both of Featherstone Winery in Niagara. Their honest appraisals of my winemaking descriptions were invaluable. Louise took special pleasure in having the opportunity to critique one of her former professors.

I would be remiss if I didn't thank the many wine industry people who came to lecture in my "labs." For ten years, I taught a wine course at the University of Guelph. Even though the course was a final-year elective in the four-year Bachelor of Commerce (Hotel & Food Administration major), it was always extremely popular. I'm quite sure it was due to the so-called labs — two-hour wine tastings held every week — rather than my skills as a lecturer. My guest speakers taught me a lot about wine and also provided me with some ongoing industry links and lasting friendships.

Finally, thanks need to go to Heather Stevens and all the team at FriesenPress. Their advice, support, and publishing expertise helped make *WineSense* into a real book — one that I can really be proud of. Thanks to everyone mentioned above, I believe this book is a good entry into the wonderful world of wines.

INTRODUCTION

"There is simply nothing else that so perfectly encapsulates physical sensation, social well-being, and aesthetic exploration at the same time." — Hugh Johnson, *Hugh Johnson's Wine Companion*

Life has many things to offer us, but for sheer enjoyment, few compare to the pleasure of drinking wine. Over the millennia, writers have verged on poetic as they tried to explain the virtues of this fermented juice. In this book, I offer my perspective on the subject — a perspective that will cut through the complexities that wine can present to all but the expert tasters. My goal is to create a clear path toward truly understanding wine and at the same time enable you to unabashedly enjoy it. I hope this book will also serve as an homage to wine; as I note throughout the book, appreciation of wine offers many rewards.

This book provides a map for discovering and grasping some of the mysteries of this wonderful beverage. It is wine simplified. You'll be able to experience this gift of nature in a deeper and more profound way. No longer will you be fearful of trying a new wine or talking about your favourite wine with more-experienced drinkers.

What is it about wine? What is the magical attraction that wine holds for so many people? Well, let's start with its appearance. At the very first glance, you see a liquid with many subtle colours and hues. Next, hold a glass of wine under your nose, and the aromas tempt you to raise it to your lips. Now allow the more tangible pleasures to unfold. The first taste is usually the fruity remnants of the grape itself. Then varying tastes and sensations come into play as wine's complexities reveal themselves. Finally, after you swallow, there is a lingering aftertaste. Perhaps a warm glow engulfs your whole body. After you take a few sips, conversations become more animated, and your better passions are released. All this from a single serving of a drink, which was conceived and nurtured in a vineyard! Wine is a muse for poets and ordinary folks alike, a great seducer and lifelong friend to many, and it's integral to some of the finest cultures on earth.

But as simple and lovely as wine may be, it can be intimidating for the novice. The names on the labels and the packaging are confusing. This situation is made worse by so-called "experts" who perpetuate an elitist approach to wine. I've attended many organized tastings led by this sort of folk. Their presentations are often so jargon-laden that I wonder what they're really trying to say. Each wine is accompanied by comments full of details and data relating to the chemical composition of the wine being tasted. They talk about "specific gravity," "brix," "phenolics," "glycerin," "tannins," "volatile acidity," "residual sugar," "malolactic fermentation," "mercaptan," and on and on. These people seem to squeeze out all the enjoyable aspects of wine. The wine itself — and all the pleasure it brings — gets lost in the analysis. It makes me want to run from the room!

In this book, I take a different approach.

The ultimate purpose of this book is to teach you how to find good and consistent styles of wine that suit your palate. Finding the right wine for the right time and the right occasion will be easy. With a virtual ocean of wine available for consumers in North America, this ability presents a huge relief when you're buying wine. You'll be equipped to sort through the vast quantity of wine and find the one you like . . . and then find others similar to your favourites.

WineSense offers you the following Three Keys to unlock the world of wine:

1. *The Basics:* This key includes fundamental knowledge about winemaking, styles and types of wine, the history of the development of wine, and some tasting guidelines. This introduction to the basics will give you a framework for understanding the rest of the book.

2. *The Grapes:* This key includes exploring the most prominent grapes used around the world as well as stories about their beginnings and their use in winemaking.

Bob Desautels

3. *The Approach:* This key explores the two main decisions a winery must make regarding its grape-blending philosophy and quality goals in creating the final product.

Mastering the Three Keys will enhance your overall enjoyment of this magical beverage and arm you with a lasting understanding of wine. You will gain WineSense, helping you continue (or embark) on an enjoyable, life-long trek exploring this elixir of fermented grapes. You'll also be able to find local wines that compare to your favourite foreign selections and receive many other rewards.

The Benefits of Drinking Wine

If you carefully read this book and diligently taste a few wines along the way, you'll eventually be seduced, I hope — seduced into a life-long quest to learn more about the wonderful world of wines and to continually indulge yourself, bathe yourself, in its many sensual pleasures. You will not be disappointed. Be somewhat moderate on the imbibing side of things, but don't be afraid of getting lost in the look, smell, and taste of wine — one sip at a time. Accept the multiple rewards that wine offers. Wine has given me some terrific comforts and lasting gifts, which I can divide into five distinct areas.

SIMPLE PLEASURE

"[Wine is] proof that God loves us and loves to see us happy." —Ben Franklin, letter to Abbé André Morellet, 1779

The one indisputable thing about wine is that it will cheer you up. The first taste of a good wine can result in an immediate delight. I can think of countless times when I tried a new wine and my first reaction was "wow!" Each subsequent mouthful leads to further confirmations that this beverage is something to behold. Then, as the wine begins to open up from exposure to air, it reveals new smells and flavours. Your appreciation ramps up sip after sip. Finally, a slight feeling of relaxation accompanied by a lift in spirits washes over you as the effects of the alcohol settle into your system. In short, wine can offer pure, unadulterated sensual pleasures.

ENHANCED SOCIAL GATHERINGS

"Wine is the most civilized thing in the world." —Ernest Hemingway, *Death in the Afternoon*

Have you ever noticed the positive contributions that a few drinks can have on a dinner party or other types of get-togethers? I have. The guests invariably become more animated, and conversations flow more easily. Thankfully, discussions often turn to those three supposedly taboo subjects: politics, sex (or is it money?), and religion. No more trivial chatter . . . the weather is just fine, thanks! It is an empirical fact, with few exceptions, that a social gathering is better when you include some wine.

My favourite toast at a dinner party is simply "Health, wealth, wisdom, and many good friends." The order and meaning of the words in my toast is quite logical. You need health to begin with . . . period. Second, the wealth I'm speaking of is not monetary; it refers to a sense of fulfillment. It means that our fundamental needs are satisfied and that we're at peace with ourselves. Third, I think wisdom is the ultimate goal of any human endeavor (and that's why it's in my toast!). And finally, one needs a lot of fine friends to share all these good things with. There you have it: My philosophy on life in just seven words. Wine just helps to bring people closer together.

HEALTH

"Wine nourishes, refreshes, and cheers. [. . .] Wherever it is lacking, medicines become necessary." —The Talmud

Countless studies show that wine is a very healthful beverage. Perhaps the most famous source for revealing wine's health benefits was a *60 Minutes* television episode titled "The French Paradox." The show highlighted the apparent contradictory fact that although many people in France have diets that are very high in saturated fats, France has very low incidences of coronary heart disease. Experts have long believed that eating a lot of fatty foods leads to a higher risk for heart disease. This connection seems to be true in many societies but not in France. Why? Some people have speculated that the one big mitigating factor is the consumption of wine, especially red wine. Each year, the French drink approximately 100 bottles of wine per capita, versus the less than 20 bottles per capita in North America. A compound called *resveratrol* seems to be responsible for many of the healthful properties of wine (the compound is higher in red wines because it comes mainly from the grape skins from which red wines get their colour). Plus, wine is high in antioxidants, which are known to have positive effects on our bodies.

Many other benefits are associated with wine consumption. Written evidence shows that people have used wine for medicinal purposes since the time of the Egyptians. Wine has long been used as an antiseptic and is said to help with digestion (I rely on it at every dinner). Interestingly, Champagne is known to be the only drink or food that a person who is seasick can digest without vomiting! Among the many other positive properties attributed to wine consumption are alleviating sleep problems, treating some nervous disorders, relieving pains, curing diarrhea, treating some cancers, and possibly controlling blood sugars. Recent evidence shows that wine may even help reduce colds and act as a memory boost for the elderly! To top it all off, there's strong evidence that moderate consumption of wine (two to four drinks a day) increases our longevity.

Sound like an amazing elixir? Not completely. Some people recognize all the good things about wine and take them as a license to increase their consumption. But as with any medicine, you can overdose with wine. When wine is enjoyed responsibly, especially with food, its allure and positive attributes are overwhelming, but overindulgence is dangerous. Excess alcohol use has been associated with various types of cancer (conversely, some studies have shown that low-level use may help prevent some cancers due to the antioxidants). The old adage holds: "Everything in moderation." Drivers, pregnant women, young people, and mentally ill folks all need to carefully monitor or avoid any use of alcohol. People with a family history of alcoholism would be wise to avoid any alcohol whatsoever.

In summary, wine can be a mixed blessing. No one expresses this better than Omar Khayyám in *The Rubáiyát:* "And much as wine has play'd the Infidel / And robb'd me of my Robe of Honour — well, / I often wonder what the Vintners buy / One half so precious as the Goods they sell."

PEACE AND FULFILLMENT

"And wine that maketh glad the heart of man . . ." —Psalm 104:15

Everyone who has had the opportunity to drink wine recognizes the feelings of warmth and satisfaction it can offer. Some refer to the sensation as that "gentle glow" one gets from just the right amount. I think the induced state of mind is best described as an amazing combination of relaxation and an uplifted spirit — a beautiful feeling of contentment. There always seems to be a point at which this feeling of peace, optimism, and fulfillment is at its peak. Unfortunately, there's always a temptation to take more wine in an attempt to ramp up this sense, or glimpse, of bliss. Drinking more is usually a big mistake, as you'll soon drift into a foggy, sometimes headachy space. Some people don't particularly care and share Lord Byron's attitude: "Let us have wine and women, mirth and laughter, / Sermons and soda-water the morning after" *(Don Juan).*

Curiously, alcohol is classified as a *depressant,* which means it inhibits the central nervous system. I think this term is a bit misleading, especially in everyday conversation. Because alcohol can put me into such a relaxed and upbeat state of mind, it certainly does not "depress" me, nor has it at any time. Using this descriptor in conjunction with wine kind of reminds me of the phrase used for a bad medical test result: "Your test results were *positive.*" Excuse me! Maybe we should find a new medical term for classifying wine, some version of my personal favourite word for wine, like *elixir.* Just a thought . . .

INSIGHT

"*In vino veritas.*" —Pliny the Elder, *Historia Naturalis*

Let's go back to some ancient wisdom as it relates to wine's final attribute. "In wine there is truth" is the English translation of Pliny's words written in Latin . . . pretty good phrase for expressing wine's most valuable — and most mysterious — gift. I, for one, have had many insights while quietly enjoying a glass or two. Sometimes I could even classify those moments as epiphanies. I can honestly say that this seemingly suddenly gained knowledge and these intuitive-like discoveries have helped me in both my

Bob Desautels

personal and professional life. Through experience, I know that wine, used judiciously, is a true treasure for which mankind should be extremely grateful. "Wine brings to light the hidden secrets of the soul," wrote Horace (a Roman poet who lived in the 1st century BC, about 100 years before Pliny).

Let us give the final word on wine to another quote from *The Odes of Horace:*

"No poems can please nor live long which are written by water drinkers. Ever since Bacchus enrolled poets, as half-crazed, amongst his Satyrs and Fauns, the sweet Muses have usually smelt of wine in the morning."

Enough said. Enjoy this book.

Cheers!

Bob Desautels, Guelph, Ontario

SECTION ONE: THE BASICS

I. THE MAKING OF WINE

"Wine is sunlight, held together by water." — *Attributed to Galileo*

The subject of winemaking often falls into a complicated technical discourse. I will attempt to avoid this trap but still equip you with some basic terms involved in producing a wine. I'll focus on the art involved along with some of the science. The goal is to give you a balanced approach, the yin and yang, to understanding this ancient beverage.

Because the grape itself has all the ingredients required to produce a fermented beverage — sugar, flavour compounds, and yeast — it's not too difficult to imagine the first experience of humans drinking wine. You can easily picture bunches of grapes stored in a hole at the back of a cave thousands of years ago. The cavemen and women return from an unsuccessful month-long hunt to find the grapes half-rotten with a pool of liquid underneath. They're starved, so they not only eat the grapes but also drink the leftover "grape juice." Instead of the sweet taste they're used to, the liquid tastes a bit sour. Yet soon they feel light-headed and then surprisingly happy and satisfied. Wine discovered! It's that simple.

What happened to our cave folks' store of grapes? When the grapes started to rot, the yeast, which is naturally found on the grape skins, came in contact with the sugar-laden juice inside the grapes. As the yeast and sugar combined, they formed alcohol and carbon dioxide gas (which escaped into the air). The end result was the creation of wine by a natural chemical reaction — a simple process, easy to duplicate and eventually practiced and refined over thousands of years.

(i) Wine Knowledge

The study of wine is formally known as *oenology*. The word comes from the Greek roots *oinos*, meaning "wine," and *logos*, meaning "knowledge."

The making of wine can be broken down into two broad categories:

- **Viticulture:** Let's start with the fundamental ingredient, the grapes. All the activities involved in planting, tending, and harvesting grapes fall under the term *viticulture*. Many people contend that viticulture is the most important part of the whole process of winemaking.

- **Vinification:** The second category covers the steps required to transform the juice of grapes into this wondrous beverage we know as wine. This area of endeavour is known as *vinification*.

Both functions involve a great deal of science and, equally, the skills of an artisan.

(ii) Viticulture: Growing Grapes

Making wine begins in the vineyard. Sound simple? It isn't. The romance of the whole process is often illustrated by pictures of beautiful rows of grape vines, a setting sun creating dramatic shadows, verdant grape leaves below which hang glistening bunches of grapes, and an attractive couple toasting one another in the forefront. Well, getting to that picture isn't easy. The farmer must consider and deal with multiple factors to arrive at the point of ripeness necessary for making wine. He or she has to start at the very beginning: in the dirt.

TERROIR

The French word for soil is *terroir*. However, in the world of wines, it means much more than the earth. It may even include the taste of the vineyard itself! There are stories of monks in the fourteenth century in Burgundy who actually tasted the dirt to determine its suitability for growing grapes. In a wine context, terroir not only includes the dirt but also refers to the location and topography of the vineyard, its aspect (e.g., north- versus south-facing), and even, to some extent, the local climatic conditions, including wind exposure.

All these factors affect the grape grower's approach to the cultivation of the vines. The farmer starts with the terroir, plants the appropriate grape variety, and nourishes the plant until the fruit reaches a correct level of ripeness. The resulting wine becomes the place, and the place becomes the wine.

CLIMATE AND SITE CLIMATE

Grape vines used for wine production grow best within temperate climate zones. In the Northern Hemisphere, all the major wine regions are within a band that runs between 30° and 50° latitude. Within this zone, you find the legendary vineyards and some of the greatest white and red wines on earth. It's worth pointing out that the more northerly regions do have one advantage: Although cooler, the days in higher latitudes are *longer* in the summer months. The grapes get less-intense but considerably more exposure to sunlight. This helps explain why even places like northern Germany are able to grow quality grapes.

White grapes tend to thrive in cooler climates better than do the red ones. This is partly due to one crucial characteristic of a good white wine: acidity. Grapes grown in more northerly areas maintain their acid levels and do not develop as much sugar. For white wine, this is a good thing as long as there's adequate ripeness, because the acidity gives the wine its desired fresh taste and keeps it in balance. Conversely, red wine grapes are best suited to slightly warmer, moderate regions. They require more warmth and often longer seasons for best results. Quality red grapes come from regions where there are not excessive heat levels. Inexpensive wines, sometimes called *bulk wines,* usually come from hotter areas like the north coast of the Mediterranean or California's central valley. These areas ensure that the grapes mature quickly — in fact, too quickly. The wines are usually too alcoholic, have minimal fruit flavours, and lack complexity; their one attraction is their low price point.

Remember that these climatic characterizations are useful mainly as a rough guideline when choosing appropriate vineyard locations. There are a number of notable exceptions to these generalizations.

An interesting phenomenon known as *site climate* is actually more important than the overall weather conditions of a particular region. Quite often, you find microclimates that are not typical of the climate of the surrounding area. For example, a hotter region like Texas can produce decent white wines, which are normally better suited to cooler, northerly climates. And in Canada's near frigid Niagara region, you'll discover excellent wines made from Chardonnay and Pinot Noir grapes. The causes of these site climates vary, but site climates are critical when determining where to establish a vineyard and choosing which grape variety to plant.

SOIL AND WATER

It has long been argued that certain soils are best suited to particular vines. Some people are convinced that Chardonnay grapes thrive in areas with high lime content, like the Champagne region in France. However, many wines that are also of excellent quality are made from Chardonnay grapes that are grown in other soils. Today's thinking seems to point more towards soil *structure* (for example, drainage) than composition as the most important factor. Yes, the mineral content has an effect, but the topography and underground physical structure seem to have a greater impact on quality and overall vine health.

Along with the climate and the soil, the third and final input for grape vines is water. The secret is finding a proper balance of just enough water. Grapes often reach optimum ripeness when the vine has to struggle to get an adequate amount of water. In lush climates, the grapes harvested have little character and result in bland and virtually one-dimensional wine. Too little water, and the vines wither, producing overly acidic, unripe fruit. The goal is to find well-drained sites, often on hillsides, where roots don't sit in water but can still find adequate amounts of moisture. (Note: In some parts of France, it is illegal to irrigate vineyards!)

GRAPEVINE MANAGEMENT

The person who's taken on the task of growing wine grapes, a *viticulturist*, has chosen a difficult and complex profession. Most of these people prefer the term *vineyard manager* or the French term *vigneron*. *Vigneron* literally means "wine grower," which is quite appropriate, as it is the health of the grape, above all, that determines the quality of the wine. The work is not only backbreaking but also very *parental* in nature. The vine must be carefully nurtured through its many stages during the growing season. The constant oversight and hard work is ultimately rewarding when the grapes finally reach maturity and are harvested.

The various stages of grape growing are as follows:

Winter pruning: The branches of the vine (known as *canes*) are cut back to limit the amount of fruit that will grow the following season. The farmer, using shears to prune the plant, does this work when the vine is dormant in the late fall or winter. He or she is cutting off the canes, leaving only the vine's main branches (see Fig. 1). The end goal is to get an optimum amount of quality fruit from each plant.

Note that there's an inverse relationship between quantity and quality when it comes to the grapes harvested per vine. Too much fruit results in poor quality, while too little results in good grapes but less revenue for the winery. Again, it's all about finding the right balance. Overall, winter pruning can be very arduous (and is usually done in cold conditions!), but it's critical to the whole process of growing good grapes.

Fig. 1

Training: The branches that grow out from the main stem or trunk of the vine must be arranged in an organized manner. The farmer uses various forms of *trellises* that support the vines during their growth. One goal is to expose the plant to the sun by spreading it out along the trellis wires (see Fig. 2); the other is to keep the branches at the right distance from the ground. In cooler regions, the vines are kept away from soil to limit exposure to frost. In warmer areas, they're trained close to the ground to take advantage of the residual heat captured by the soil or rocks during the daytime — the vines are then kept warm during the evening as that heat is gently released.

Fig. 2

Fruit emergence: As the weather begins to warm in the fields, the vine goes through a number of changes. Starting with a general coming to life, the plant soon produces buds — a critical stage in the growth cycle. At this point, the plant is especially susceptible to frost, which could destroy the crop (most commonly a problem in northern climates). Next come the flowering stage and the emergence of the tiny grape bunches. As the bunches start to ripen, the small grapes begin to develop colour — a period known as *veraison.* The grapes then change from a hard to soft consistency.

Summer thinning: Two main types of vineyard thinning occur during the warmest part of the season. First, the leaves of the vine are managed carefully. They're removed at certain points in the season to allow sunlight to reach the grapes and also to improve ventilation; both factors will improve ripeness, limit grape rot, and stop certain mildews from damaging the fruit. This practice is called *canopy management*. The second type of thinning is not done universally. In some vineyards, in an effort to maximize quality, the farmer removes bunches of grapes from each vine. This is called *crop thinning*. After a quick study of the plant, guided by years of experience, the farmer shears off the poorer quality bunches to ensure that the remaining bunches will thrive. I once heard a well-known vigneron say, "It's not the fruit on the vine that determines the quality. It's the grapes left on the ground."

Farming philosophy: I've always been amazed at the amount of chemicals used in modern grape growing. Potassium, phosphates, sulphur, nitrogen, pesticides, herbicides, etc., etc. — all employed to control the various pests and elements affecting the vines. The poor farmer and vineyard workers are constantly exposed to any number of these potentially harmful substances. And we, as consumers, must surely be getting some residual chemicals in our wine. A lot of these products were specifically introduced to (a) control disease, especially various forms of fungi and (b) ensure consistent production. Many wineries are returning to traditional farming and more natural approaches, which can be more labour-intensive. They still use some chemicals, like sulphur which is natural, but limit their use. Today we call this approach natural or *organic*. I, for one, am all for it (throw in some *biodynamic* too!).

The harvest: The last stage in vineyard farming activities is the annual harvest. The farmer must pay special attention to the condition of the fruit and determine the right moment to bring in the grapes. Limiting any significant damage caused by careless harvesting is imperative. After all, the farmer has just spent months and months nurturing the vines — guarding against various diseases, keeping birds and other animals away from the grapes, and patiently waiting until the fruit is sufficiently ripened. Damaged fruit will be susceptible to various infections during the time from harvesting to pressing, which could affect the quality of the finished wine. However, most wineries ensure that picked grapes are taken to the winery quickly, before any problems arise.

One critical function for the farmer is determining when to pick the grapes — a decision governed by experience and exacting science. A skilled farmer can tell whether the grapes are ready for harvesting merely by tasting one. To produce a good wine, the goal is to have a perfect balance between sugars and acids. Science can play a role here too, giving an objective measure of ripeness. A small device, a *refractometer*, is an invaluable tool for the farmer. Used in the vineyard, it measures the exact levels of sugar in grapes.

Basically, there are two methods for picking grapes: manual and mechanical (see Figs. 3 and 4). The first technique goes back to the very beginning of winemaking. Bunches are snipped from the vine, placed in baskets, and taken into the winery. Today, many of the best wineries still pick by hand to ensure the grapes are not bruised and that quality is maintained until pressing. The other technique involves large mechanical harvesters. These machines essentially straddle a row of vines and shake the plants so that the grape bunches fall into containers attached to the harvester. The savings in labour — and associated costs — are enormous. The logic for this technique is best summed up in one word: *efficiency.* For many wines, mechanical harvesting is a fine method for bringing in the harvest, and it reduces the timeline from picking to pressing. However, the high-quality wineries still use experienced pickers.

Fig. 3

Fig. 4

The work of viticulture is now complete. As you can imagine, there's a lot more to the subject, but you know the basics.

QUALITY VERSUS QUANTITY

At this point, it's perhaps worth reviewing grape growing with a sharper eye on *quality*. The guiding principal is that fewer grape bunches per vine will yield better quality. The basic premise is that more nutrients flow into the limited quantities of fruit. Winemakers often make the vine struggle a little so that the roots of the vine go deeper to get water and find more complex minerals. In parts of the world, irrigation is forbidden or at least strictly limited under the belief that limiting water will make a sturdier plant that will be more resistant to droughts and less susceptible to disease. Experience shows that these vines also produce finer grapes with more complex and concentrated fruit flavours.

The quest for quality by reducing quantity also has some limiting factors . . . for example, cost! If you're interested in inexpensive bulk wines where taste is less important and low cost is paramount, then you maximize the harvest. In these cases, pruning of the grape branches is designed to maximize crop load. There's little or no thinning of bunches in midsummer. Irrigation and fertilizers are used generously, and the crop is plentiful. The result is a lot of grapes, sometimes more than 20 tons per acre. By comparison, in the high quality wine regions, harvests are sometimes less than 4 tons per acre — less than a fifth of that in the high-producing areas!

The tradeoff in the two approaches might be more accurately called *quality versus cost*. As in many other human endeavours, the difference lies in the attention to detail and the hard work. All that attention and time spent pruning, thinning, hand-picking, etc. — it all adds up in terms of cost and effort, but it's most often worth it. We'll discuss this whole subject more fully in Section Three of the book.

(iii) Vinification: Turning Grapes into Wine

Actual winemaking is an activity that gives great pleasure, and it's not nearly as arduous as growing grapes. I'm speaking from experience. For years I made wine with a group of friends. We began by monitoring the progress of the year's crop, or *vintage*. Sometimes this involved a visit to the winery and time spent watching the farmers at work, getting their opinions on the season. Then came our familiar ritual. It began with a call from the winery telling us our grapes were ready. We'd drive down to the vineyard and load up our pick-up truck, toast the vineyard owner with a glass of his wine, and head back to one of our garages. The grapes were put into a de-stemmer and crushed (if they were white grapes, we'd press right away . . . more about that later in this section). Then we'd add a bit of sulphur (metabisulfate is a common form of sulphur dioxide) to stabilize the juice and kill undesirable microbes. Next, the yeast would be added to start the fermentation. A week or so later, we had wine to age and store. When we got around to bottling, we'd open a wine made with same grape from the previous year and drink it abundantly, congratulating ourselves on this and last year's crop. The final step was not necessary but definitely added to the pleasure!

Now let's go through some of the winemaking steps in more detail.

DE-STEMMING AND CRUSHING

Once the harvest is complete, the grape bunches are quickly transported to the winery. The better wineries are very careful about the individual grapes' condition and examine them carefully as they arrive. Excessive bruising can lead to infections of bacteria or fungi and even cause some premature fermentation. Once the fruit is deemed fit for the next stage in winemaking, the grapes are immediately put into a de-stemmer, leaving mainly the grapes and some escaped juice (referred to as *free run* juice). Everything then continues on its journey as it's put into a crusher (most wineries de-stem and crush simultaneously), which further breaks down the grape bunches into a slushy concoction of juice, pits, and grape flesh. At this point, you need to decide whether you're going to make white or red wine.

WHITE OR RED?

After the crush, the winemaker must make a choice about the colour of the wine — but only if the grapes are red. Virtually all grape juice is white, regardless of grape colour. In fact, the best known wine in the world, Champagne, is often made from two red grape varieties and only one white variety. The colour of wine comes from the skins. Therefore, you can make white wine from red grapes, but you cannot make red wine from white grapes.

Depending on the wine being made, the crushed grapes are either pressed right away (white wine) or left to ferment on the skins (red wine). This is a bit of an oversimplification, because sometimes white grape skins are briefly left in contact with the juice to get more flavour and complexity in white wines — and even after fermentation, the sediment (mainly dead yeast cells called *lees*) is sometimes left on the bottom of the storage vat for the same reason: to add complexity. In general, however, whites are pressed quickly to produce a lighter, more refreshing drink. The skins tend to not only add colour but also introduce astringency, which is more appropriate in "bigger," more intensely flavoured red wines.

PRESSING AND JUICE CLARIFICATION

There are two main methods for pressing grapes. The first kind, the *basket press* (see Fig. 5), has not changed for over a thousand years. Grapes are placed in a round barrel or container that has small openings along its sides. A circular wooden plate is screwed down the "basket," causing the juice to flow down the sides of the container before being funneled into a catch basin.

The second kind of press is called a *bladder press* (see Figs. 6 and 7). These can be either vertical or horizontal containers, usually stainless steel, which have an inflatable bladder in the middle. The press is loaded with the grapes, and as the bladder expands against the walls, the juice is released through openings on the sides. The whole contraption slowly rotates to keep the pressure constant and stop the grapes from collecting unevenly. The advantage of the bladder press is that the grapes are crushed more gently so that fewer extracts from the skins and pits, which can be bitter, end up in the juice.

Fig. 5

In the case of white wine, the pressing occurs before fermentation. After fermentation, the juice is *clarified*. Either the juice is allowed to settle in large tanks before the clear juice is drawn off (a process known as *racking*), or the juice is put through a centrifuge to clarify the liquid. For red wines, the pressing and clarification steps are done *after* fermentation. Also note that with red wine, especially quality reds, the use of the centrifuge is less common because it removes too many elements, adversely affecting complexity. All reds go through the racking process, often a number of times before bottling, whereby the wine is drawn off the top, leaving the sediment on the bottom of the tank or wooden barrel.

FERMENTATION

Although the actual process of fermentation has been observed since the beginning of time — wine time, anyway — a comprehensive understanding of that process did not emerge until the 19th and 20th centuries. It was one of those great mysteries. The word itself means "boiling" or "bubbling" because of the heat it generates (the bubbling action you can observe as it happens). In the 1850s, the Frenchman Louis Pasteur discovered the connection between yeasts and sugar that explained the fermentation process and its importance in producing wine. He discovered that the yeast essentially attacked or consumed the sugars. This critical interaction created heat and effervescence. Eventually, the sugars in the grape juice were converted to alcohol and carbon dioxide.

Ambient versus cultural yeast: Wine is a natural phenomenon, and its "discovery" is no surprise. Yeasts are found (almost) everywhere — on the grapes (sometimes you can see them in a form of a white dusting or *bloom* on the skins), in the air, in the winery's cellar . . . or in a cavemen's home! These wild, naturally occurring yeasts, referred to as *ambient yeast*, will turn any pile of old rotting grapes and juice into wine. Sometimes they're simply allowed to do their work in the winery under highly controlled conditions. This natural, hands-off practice is still followed in some of the best wineries around the world.

Most commonly, however, these ambient yeasts are destroyed by adding sulphur compounds. Winemakers employ this approach for two main reasons. First, wild yeasts are unpredictable and can cause odd flavours or actual spoilage of the wine. Second, yeast itself contributes some flavours, and most winemakers want to control that taste factor. Many yeast strains are available, and many are best-suited to certain grapes and/or certain styles of wine. In fact, a different strain used on the same grape, in the same winery, will cause varying flavour.

Red wine fermentation: It is worth repeating that the major difference between making red versus white wine is that reds are fermented *with* the grape skins of red grapes. Again, this is not as simple as it sounds. As the red wine goes through the fermentation in large tanks, all the skins (and the stems if left in the batch) tend to rise up to the top, forming a *cap*. This creates a problem because it minimizes the skin contact, thereby limiting extraction of colour and other elements, like tannins, which also contribute to red wine's character.

Winemakers take one of three traditional strategies to combat the formation of the cap during fermentation. The first approach is to simply push the skins down into the wine. This is known as *punching down* (or *pigeage* in French), and I'm sure it's been practiced for hundreds of years. Another common process involves the use of a pump. The clear wine from the bottom of the fermentation tank is pumped over the top of the cap. This allows the wine to soak through the skins and pick up colour and added complexity. This process is very similar to a coffee percolator (the French call this *remontage*). The final strategy is to use fermenting tanks that rotate on a horizontal plane (i.e., the wine tank lies on its side and turns like a cement mixer). This technique is used in more modern and often large wineries.

Malolactic fermentation: This secondary fermentation is not a result of yeast acting on grape juice; rather, it is a bacterial fermentation that occurs in wine. It is a desirable step in the making of certain wines, especially most reds.

During malolactic fermentation, the naturally occurring malic acid is converted to lactic acid. Malic tends to be the tarter tasting of the two acids, while lactic acid, which is the same acid found in dairy products, is a softer tasting element — you often get a "yogurt" aroma along with a buttery aftertaste in wines that have gone through malolactic fermentation. In general, this fermentation is encouraged in red wines because they don't need as much acidity in the final product. Reds have other astringent characteristics to give the wine complexity and substance. Conversely, in whites, the acidic character of the malic acid adds to its freshness and tends to highlight its fruit flavours. Therefore, most whites are *prevented* from undergoing malolactic fermentation.

PREPARING FOR BOTTLING: CLARIFICATION

Fermentation is complete when all the sugars have been converted to alcohol.[1] The wine has lost all its sweetness and is now called a *dry* wine. I'm not sure of the origin of the term *dry*, but it is widely used to define wines with little or no residual sugar. (The majority of table wines, common everyday wines, are dry.) At this point, the wine must be prepared for bottling by removing suspended particles and clarifying the end product.

Racking: Due to gravity, wine will naturally clarify as any sediment drops to the bottom of whichever container the wine is being stored in. The clear wine is drawn or racked off this sediment. After fermentation, this sediment consists mainly of dead yeast cells and minute pieces of skins known as the *lees.* The racking process is often done a number of times as the wine ages. Particulate matter continues to form and precipitate out as the wine matures. Higher quality wines, especially reds, are often left in contact with these minute particles for longer periods. You'll sometimes find sediment in a bottle of older quality wine when you pour it out at home — don't worry, as it's harmless. The wine has been stabilized in an aging barrel.

Fining and filtering: The winemaker uses two other tools to further clarify the product. The first is known as *fining,* whereby a substance is added to the wine to draw out suspended particles. This substance can include egg white, bentonite (clay), or even isinglass (from fish bladders!). Fining substances attract particles, which in turn coagulate and fall to the bottom of the aging tank or barrel. This technique is often used for the best wines, as it is viewed as a more natural and gentler process.

The other approach to clarifying wine is through the use of filtering machines. These come in varying sizes and levels of efficiency. In large, modern wineries, they are high-tech and are designed to remove any suspended particles in the wine. The wine is almost "sterilized," which ensures greater stability but at a cost: flavour.

....

1   *Unfermented sugar:* Understand that yeast will stop working when the wine reaches high levels of alcohol (approximately 15%). The yeast produces the alcohol, which in turn kills it off! Later in the book, I discuss styles of wine that are purposely made to be sweet because of residual sugars.

Realize that consumers want a clear wine in their glasses. The winemaker knows this and will use a number of appropriate processes to make that happen. However, winemakers concerned with quality keep their interventions to a minimum. They believe the wine needs to be "whole" and include most of the natural elements; many winemakers will not filter their wines at all. The danger in this approach is that the wine will throw sediment in the bottle (often a sign of a better wine and therefore a good thing!) or, worse, the wine will start to ferment again due to some live yeasts or bacteria that weren't removed. It's often a tradeoff between a complex taste and quality on the one hand and uniformity and stability on the other.

BOTTLING AND AGING

The title of this section could have just as easily been "Aging and Bottling," because some wines are aged before the bottling, while others do the majority of aging in the bottle. Nevertheless, I will talk about the main factors that allow wines to age well. The process of filling bottles is the final *active* involvement in the winemaker's catalogue of things to do in the winery. The intriguing alchemy of turning juice into wine, vinification, is complete. But the techniques used all the way up to bottling in the winemaking process will give the wine the essential characteristics that determine its age worthiness.

Aging factors: Many factors affect a wine's aging potential. I will discuss only the four most important factors that affect the longevity of a wine. First is the alcohol level. Generally speaking, the greater the amount of alcohol, the more potential the wine has to stay healthy in a bottle. However, alcohol does not contribute significantly to any improvement in a wine's taste as it ages.

Second, tannins from the grape skins and/or wood barrels help a wine to age. These substances act as a preservative and have an astringent taste (if you've ever had very strong black tea, then you've experienced the taste of tannins). In a very young wine that has been made to maximize tannin content, you find the taste quite bitter. In fact, your mouth puckers and the wine is often almost undrinkable. Yet after a few years of aging, the taste mellows because the tannins combine with other elements in the wine to produce new, softer tastes and intriguing smells. An almost magical meta-morphosis takes place with certain great wines. They undergo changes quite dramatically during their aging. When the wines are consumed after aging for a few years, wine drinkers sometimes use the strangest, almost unbeliev-able descriptions of a good vintage wine, such as "smells of violets"; "it has immense complexity"; "vegetative barnyard aromas are mixed with black-berries and tar"; "cascading flavours of dried fruits and velvety mouth feel"; and "a cat pee bouquet and gooseberry taste." More than any other factor, tannins, especially in reds, help a wine age and contribute to its transforma-tion. The winemaker decides how much tannin will be extracted into the wine by controlling skin and oak barrel contact during fermentation and storage. There must always be a fine balance between fruit character and tannin levels. If tannin is the only prominent element in the wine, the result will be a one-dimensional, astringent wine.

The third factor affecting aging is the acid level. We all know that acid preserves food — think of pickles. And in wine, acid has the same prop-erties. Acid is a natural component of the grape, and it survives primary fermentation. Most reds go through a secondary malolactic fermentation, which reduces the acidity in a wine. However, in white wines, the acid gives the wine its freshness and tartness. It is the element that gives whites their structure. Think of acid as the backbone of white wine (while reds depend more on the tannins for structure). Without the acids, the wine will taste flaccid (often described as "flabby") and uninteresting. In some parts of the world, winemakers add acidity, normally through tartaric acid, to enhance the final product and improve its stability.

The final essential factor required for aging a wine is its sugar level. Again, everyone understands the value of sugar in keeping foods. Its presence in wine has the same effect. Some of the best-preserved wines are dessert wines like late-harvest wines or Sauternes, which have high levels of sweet-ness. These wines are made from grapes that have high levels of natural sugars that remain even after fermentation; the yeast cannot fully consume all the sugar. These residual sugars give the wine aging potential. Sometimes extra-sweet wines are combined or "fortified" with higher levels of alcohol, as with some Ports and Madeiras, resulting in wines that can age for over a hundred years!

One final note on aging wines: 95% of wines are not meant to "live" beyond a few years. For various reasons, including climate, grape varieties, vineyard practices, and winemaking techniques, most wines are meant to be drunk young. As they age, they will not have sufficient amounts or balance of any one, or more, of the four aging factors of *alcohol, tannin, acid,* or *sugar.*

Pre-bottling aging: The winemaker has many tools to influence a wine's health and longevity. For example, to *increase* tannins, practices include leaving the wine in contact with the skins for longer periods and/or fermenting and storing in oak barrels (which also contain tannin). In white wine production, the goal is mainly to *preserve* the acidity to maintain a crisp, fresh flavour profile. The winemaker fines and filters the wine immediately after fermentation and then cools it quickly. These activities all ensure the whites avoid malolactic fermentation[2], which would soften the acidity, and help maintain fresh fruit flavours so integral to most whites. Many other tools are at the disposal of the winemaker to affect the wine's aging potential. We'll just look at one more, the use of oak.

Oak barrels were originally used as containers for shipping wine. They replaced amphoras[3] of the ancient world. Oak was more durable and therefore easier to transport. Over time, people noticed that the barrels affected the wine's taste, often in a positive way.

Today oak is used in some fermenting tanks (stainless steel is more common) and as barrels for aging and stabilizing the wine, mostly reds, prior to bottling. Its influence on the wine is determined by a number of factors. One primary consideration is the condition of the wood. New barrels will impart more flavour, and the wine will also extract more tannin from the oak itself. The type of wood also plays a part. For example, American oak barrels give a stronger taste and vanilla-like aromas to the wine, compared to oak from the old forests in France.

....

2   *Chardonnay exception:* This white wine grape is often allowed to go through malolactic fermentation and is sometimes aged in oak barrels as well.

3   *Amphora:* These containers, or "vessels," were usually ceramic and were used in Greek and Roman times to transport foods and beverages (see Fig. 9).

How long the wine is left in the oak is another important consideration. The effect is two-sided. On one hand, with longer exposure, the wine will extract more tannin, taste more astringent, and have greater aging potential. Conversely, wine barrels allow a wine to breathe — that is, to be exposed to oxygen — which can be both a positive and negative thing. Oxygen is normally an enemy of wine, but in small quantities over time, oxygen allows a wine to soften and mature more quickly. It is up to the winemaker to find the right balance. Unfortunately, many winemakers go overboard and end up with an over-oaked product where the fruit taste is lost. Great wines find the midpoint where the fruitiness dances in the oakiness and then change in fascinating ways as the wine ages.

Bottling considerations: The use of glass bottles dates back many centuries, as the bottle was found to be an excellent container. Glass does not affect the wine. It is inert. However, there is a problem with bottles: They need an opening to get the wine in and out! This opening exposes the wine to oxygen. Cork has traditionally been used as the stopper for the bottle. This wood, mainly found in Portugal and Spain, is ideal because it keeps oxygen from getting to the wine, thereby ensuring the wine is safe for a *relatively* long time. After 20 to 30 years (sometimes much earlier!), all cork will begin to deteriorate, and as a result, so will the wine due to oxidization. Today you see two other types of stoppers: synthetic plastic "corks" and metal screw caps. The advantage is that no wine is spoiled. Some traditionalists refuse to use anything other than cork. They claim a cork allows the wine to breathe in a very slow, controlled manner. They believe this is important to maturation. This theory is widely disputed, and the two modern closures are becoming more and more common.

WHITE WINE MAKING

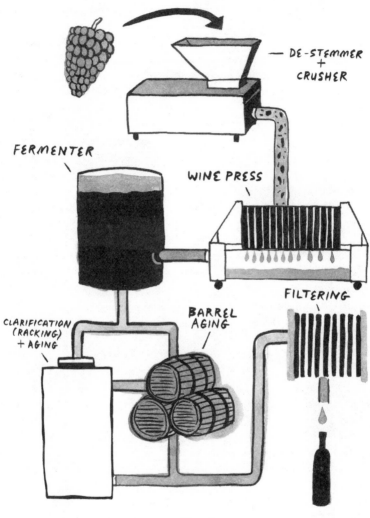

— DE-STEMMER
+
CRUSHER

FERMENTER

WINE PRESS

CLARIFICATION
(RACKING)
+ AGING

BARREL
AGING

FILTERING

Fig. 6

RED WINE MAKING

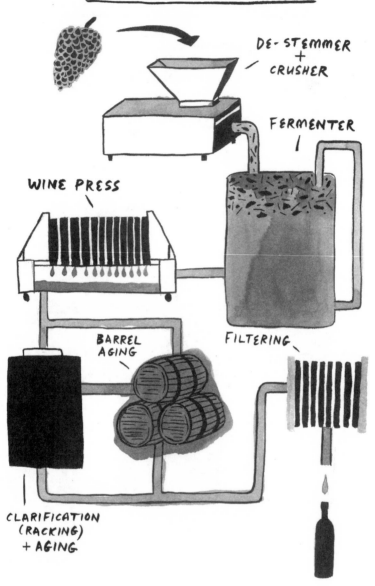

DE-STEMMER
+
CRUSHER

FERMENTER

WINE PRESS

FILTERING

BARREL
AGING

CLARIFICATION
(RACKING)
+ AGING

Fig. 7

VINTAGES

I hope you've enjoyed the first part of Section One, "The Making of Wine." Winemaking can be somewhat complex, but I've tried to keep the explanation simple and therefore avoided many technical terms as well as more-involved winemaking procedures. The goal is to cover the subject of wine-making sufficiently so you can fully integrate the Three Keys. I'll discuss one more term, *vintage,* prior to moving on to the more palatable parts of the book.

Vintage simply refers to a year . . . usually. If a wine label has a date on it, the *grapes were grown in that year.* The word *vintage* does not describe the quality of the wine as many people seem to think. "Ah, it was a vintage wine we were served" is the type of comment I've often overheard. The confusion comes from two sources: climate and Champagne.

The growing seasons in any grape-growing area obviously vary from year to year. Some are more suitable for growing ripe fruit that has the right mix of acids and sugars. The result of a good season is a better wine, so many people say, "It was a vintage year" when what they really mean to say is, "That year's vintage was excellent for growing the type of grapes that make great wine." It would make it easier for all novice wine lovers — budding oenophiles — if people described a particularly good wine as coming from a "great vintage."

Blame Champagne and Ports for confusing things even further. Both wines are usually made by blending wines from different years (i.e., different vintages). However, every so often, each area will *declare* a vintage year (sound redundant?) when the grapes are determined to be of excellent quality. In those years, instead of making only the blended wines, the producers make a special wine using *only* that year's grapes. This is usually a better and more expensive wine and will have the year stamped on the label. I hope this ends the confusion.

II. WINE TYPES AND STYLES

Never apologise for, or be ashamed of, your own taste in
wine. Preferences for wines vary just as much as those for
art or music." —Hubrecht Duijker, Dutch wine writer

My own introduction to wine was similar to that of many other North Americans. Most of us had been raised on juice and pop, a habit that resulted in a taste for sweet beverages. I first tasted wine when I was ten years old at my parents' dinner table. The wine was red, no doubt dry, and did not go down well! I couldn't understand the fascination. I remember that it tasted vaguely like unripe, dried berries with a strange, bitter aftertaste that lingered in my throat. For the next few years I avoided wine, mostly, and instead tried to develop a taste for beer. Beer didn't fit the sweet beverage preferences of my youth, either, but with peer pressure as my motivator, I was happily quaffing pints of draft in my late teens. Then came university.

At one of the first parties in university residence, I was introduced to a sparkling semi-sweet Portuguese rosé. Not bad, I thought. Quite sophisticated, I also thought. It didn't take long before I was sharing California Chardonnays with friends and starting to appreciate a more multidimensional flavour — fruity upfront, tingling and buttery on the sides of my mouth, lingering tastes of pears and remarkable overall fresh, clean feeling on my palate. Not long after, I ventured into reds and was soon laughing about my bubbling rosé days.

So that was my introduction to the world of what is called *table wines,* consisting of white, rosé, and red wine. Table wines are just one of the four major *types* of wines that I'll discuss in this part of the book. Within each type are varying *styles.* For further clarification, think of wine types as distinct classifications and wine styles as differing characteristics within each type grouping.

(i) Table Wines: White, Rosé, and Red

The vast majority of wines made in the world are table wines.[4] They range from everyday, inexpensive wines used as an accompaniment for normal meals and activities to expensive wines with incredible attributes brought out only for special occasions. Most are made for consumption within a year or two of the vintage. In fact, very few will improve or develop with aging (like a lot of us!). Table wines can be broken down into three broad colours, or subtypes. Each subtype also has varying styles.

WHITE WINE

Whites are normally delicate or light tasting, fruity, and fresh on the palate. The main exception to this rule is certain Chardonnay wines, which are frequently exposed to the skins and lees during and after fermentation. These robust whites are often aged in oak barrels, which add even more flavours, including tannins. I've seen these wines mistaken for reds during tastings when people were blindfolded . . . it's quite amazing!

White wines are also light in another way. They almost always have less alcohol content than red wines. The normal range for a white wine is 10–12% alcohol by volume. Some German wines, especially from the Mosel region, have as little as 7% alcohol by volume. In general, whites are delicious and great to drink simply as an aperitif. Because the alcohol "punch" is minimal, compared to the blockbuster 15%-alcohol-by-volume reds, it's not often necessary to have food with many white wines.

Let's now look at the main styles of white wines. Remember the stages of the winemaking process: The style of a wine (red, rosé, or white) is a result of the terroir, overall climate, growing practices, vintage, and winemaking

---

4   *Table wines:* This term describes wines that are usually between 10% and 14% alcohol.

techniques of the winery. It also depends on the grape variety. We'll discuss this last factor in Section Two, titled "Grapes."

Plain, light, neutral whites: This style is sometimes called "winey." The description is apt, but I can't say exactly why. I think it's because although the wine is *technically* wine (i.e., it contains alcohol), it has little distinctive flavour. Or it could be that the wine is so bad tasting that we tend to whine about it!

Unfortunately, a lot of bulk wines fall into this category or style. When making dry, neutral whites, the winery often uses grape varieties that do not have much character to begin with. Or conversely, they've used a classic quality grape but allowed it to flourish on the vine instead of *cropping* the bunches to reduce the harvest; cropping yields more-concentrated grapes, which are the raw material for great wine. Regardless of the reason, this style is at least inexpensive and good to use as a quaffing wine for certain occasions . . . or if you're desperate.

Fruity and flowery: This style of white is one to drink all by itself. It doesn't need food to be enjoyed. These wines have aromas that will probably remind you of different fruits and flowers. The initial taste is sweet, but the aftertaste, or *finish,* is often dry due to the high acid levels in some whites. The great wines in this style grouping display a perfect balance between the sweet fruit and a pleasant "bracing" freshness from the subtle acidity.

Typical examples of this style usually hail from cooler grape-growing areas. Think Germany, northern France, Washington and Oregon, New York State/Niagara, New Zealand, and many coastal and/or mountainous regions.

Complex and full-bodied: These are the whites that often go best with food. They tend to be drier on your palate, and the aromas are more exotic. There's a less-direct association to a particular grape, and other fruit smells tend to predominate. Many of these whites are made from fully ripe grapes, which often mean lower acidity, and as a result draw a lot of their flavours from the fermentation and aging process. This style can also be enhanced with greater skin contact and oak barrel storage prior to bottling.

Many of the great examples of complex whites come from moderate climates like Napa Valley, Bordeaux and Burgundy in France, northern Italy, and Spain. The majority are made from the ubiquitous Chardonnay grape, but I think that Sauvignon Blanc (and blends of other grapes with the Sauvignon Blanc) and older aged Rieslings also have a place among some of the finest wines in this style grouping.

I should mention a little more about the exotic smells I referred to earlier. Some might even call these smells weird. For example, how can the aroma of cat pee be a good thing in a wine? Well, think about that smell and then try a well-made, older Sauvignon Blanc from New Zealand or the Loire Valley in France. There are many other examples of weird and good smells, such as the barnyard smells from the red wine grape Pinot Noir. And what about old cheeses? The smell is rarely indicative of the actual taste.

Grapes are like caterpillars in that they go through a metamorphosis during the fermentation and aging process. Various compounds, known as *esters,* form at some point during the aging process and as a byproduct produce different smells. The same thing happens as other elements combine in the wine to produce interesting tastes. The result is a more complex, and interesting, wine than the neutral white style.

ROSÉ WINES

Rosé wines have been given a bad rap over the last few decades. This is largely due to a lot of lousy examples in the marketplace — think cheap pink Zinfandels. Rosé can be delightful. In fact, in France (especially in the south), rosé wines are sometimes more popular than whites and reds!

The proper way to make this type of wine is to allow grape juice to soak, or *macerate,* on the red skins for a short time to extract just a little colour. Then with a gentle pressing, the *pinkish* juice is collected and fermented.[5] The result is a beautifully coloured wine with some attributes of lighter whites and some complexities found in reds. Find the right balance, and you have a lovely "sipper" for a hot day or a versatile food-friendly wine.

....

5   *Saignée:* The French perfected this method of making rosé wine and it is referred to as saignée.

Dry rosé: This style is made by ensuring all the sugar is converted to alcohol during fermentation. This wine requires better grapes because there will not be much sugar to contribute to the flavour . . . or to mask the fact that there may be less flavour in the grape. The flavour comes from all the other elements in the grape, including the colour and tannins derived from the skin contact. If the wine is allowed to sit a bit longer with the red grape skins, then there will be more colour, often bordering on orange tints, and a more-complex taste.

Some of the best examples of this rosé style come from France and Spain and increasingly from the so-called New World of California, Australia, and Chile. Also look to cooler climates like Canada and the northern U.S. for some delightful lighter styles ideal for drinking as an aperitif.

Sweet and fruity: For some people, this style of wine has little place at the table or even in the glass! However, I feel this kind of rosé, with lots of fruitiness and some residual sugars, has its place in the cellar. These wines can be a perfect match for many desserts, like a berry tart or an apple strudel. Look to Portugal and California for some pleasant rosés in this style.

RED WINE

Whereas white wines are often enjoyed alone, red wines almost cry out for food accompaniment. Of course, there are exceptions, but in general, red is made for the dinner table. I just consider red wine a food that often takes a role similar to a sauce.

The other factor that impacts red wine's place in our patterns of consumption is alcohol content. Most red wines average 12% or more. In fact, recent trends have seen that percentage climb consistently nearer to 14%. This higher alcohol content is a result of better vineyard practices in some cases, and it also comes from manipulation of the vinification process.[6] There are various reasons for this trend in the world of wines, and not all of them are positive. One common opinion is that certain influential wine critics have overly tilted the measure of quality towards big, intensely concentrated

....

6   *High alcohol wine:* One way to increase a wine's alcohol content is to add sugar to the juice before fermentation —a practice known as *chaptalization.*

wines, particularly reds. A lot of the subtlety of red wines has been sacrificed as a result.

Why is food so important when drinking red wine? In the presence of food, alcohol is absorbed into your bloodstream at an appreciably slower rate. As a result, you don't get that hit to the brain from the booze. Instead, you get a pleasant transition into that very enjoyable state of satiation from the food and the warm, satisfying effect of alcohol. As the food minimizes alcohol impact by slowing down absorption, the wine helps the body digest the food. The combination is a good prescription for health in my opinion. I cannot imagine dinner without wine; I even have trouble thinking about the proposition! Just remember: Everything in moderation.

Plain, simple "bulk" reds: As with whites, this basic style of red wine comprises the vast majority of reds produced. However, I find that unlike neutral white wines, this style of red is frequently very acceptable. Like most people, I have a limited wine budget (I'll admit, however, that my limit is larger than average!). That aside, I am usually satisfied with a bottle of bulk red at dinner. Perhaps it's because a red wine tends to have some character, and if it is well made, then it comes across as a nice complement to many everyday red meats and pastas. The bulk red character is in opposition to the winey, bulk whites that are offensive because of their complete lack of flavour. Like much fast food, the complete neutralization of any strong flavours and the addition of some sugar has an inverse effect: Its very inoffensiveness makes it exceedingly offensive! (Now I think everyone is clear about my feelings regarding fast food!)

Young, fruity smooth: This style of red wine lends itself to being drunk by itself, without food, as most whites do. Red wine of this kind is usually made from grapes that contain an inherently lighter fruit flavour. The fruit comes from regions where the grape ripens easily and large yields are allowed to develop on the vine while keeping the acid levels low. In the winery, the grapes have less contact with the skin, and the fermentation is often done at cooler temperatures (like most whites).

My favorite reds of this style are best illustrated by the wine of Beaujolais located in southern Burgundy, France. Beaujolais wines lean toward light fruit nectar but have a nice complexity to round out the flavours. This wine is a perfect wine for a barbeque. You can drink it as a hamburger sizzles on the grill and then drink more as you eat the burger smothered in condiments and cheddar cheese. Other great examples of this style hail from California, Chile, northern regions of most European countries, Australia, and New Zealand. Look for grapes like the Merlot, Pinot Noir, Shiraz, Gamay, Zinfandel, and Cabernet Franc.

Mature, complex, balanced: Here is the red wine style that can set the world on fire and that sends both the poets and the wine experts into moments of rapture. With these wines, the winemaker has taken the perfectly ripened grape and then carefully nurtured the vinification process so that a truly great wine emerges. Often, a few years of barrel and bottle aging are required for the wine to reach its peak. Believe me — it is worth the wait.

The central goal of the winemaker is to achieve a balance among fruit, acidity, oak flavours, and tannins along with the many other elements from the grape itself. The end result, if achieved, is a subtle collection of aromas, mouth feel, flavours, and aftertaste (or *finish*). No single characteristic dominates. The oak melds into the fruit, the aromas are distinct and numerous, the wine tastes like an integrated whole with multifaceted components. I suppose you've gathered by now that I like this style of red. It is the epitome of great wine.

Traditionally, the very best of this style has come from the two French regions of Burgundy and Bordeaux. Other old-world regions include Spain's Rioja and Italy's Tuscany regions. Today you'll find tremendous expressions of this style in Sonoma and Napa Valley, Oregon and Washington State, Chile, New Zealand, and Australia. The best age for these wines is usually between 5 and 10 years. Look for 12–13% alcohol.

Concentrated, powerful reds: This style of wine has become extremely popular among many wine drinkers in the past few years. It is part of a trend that depends on better vine management by limiting the amount of grape bunches during the growing season. The result is more nutrients going into fewer grapes, creating a greater concentration of flavours. As I mentioned previously, the other reason for this trend is that the public perceives a connection between a wine's quality and the intensity of flavours.

This perception of quality also extends to wine with high alcohol levels. Historically, during hot summers and dry autumns, the grapes would naturally (i.e., without pruning or thinning) ripen more than normal, and with proper aging, the resulting wine would be very powerful. Today the same result can often be achieved even in mediocre seasons by aggressively limiting the crop. The loss in grape yields — and therefore quantity of wine produced — is made up by higher prices due to the increased demand for this style of red wine.

In my opinion, despite the concentration of flavours, this wine often tastes forced and artificial (especially when drunk young), and with the exception of naturally occurring great vintages, I don't think this style of wine is particularly compelling. It reminds me of the saying (paraphrasing here) "using a sledgehammer to drive in a little nail." The personality of the wine is lost in an overpowering concentration of fruit extract, sometimes accompanied by too much oak, and enveloped in ridiculously high, nose-burning levels of alcohol (14–15%!). Yet these wines are praised by some of the most influential wine critics, most notably Robert Parker, and are therefore avidly purchased at astounding prices by the gullible wine-buying public. Remember to trust your palate and not feel intimidated by experts' comments!

Overall, these sorts of powerful reds go wonderfully with beef, red game meats like venison, strong cheeses, and spicy pastas (we'll talk more about matching food and wine in the book's appendix). Some of the best examples of this style come from California, Australia, northern Italy's Piedmonte, and the Rhone Valley region in France.

(ii) Dessert Wines

The word *dessert* can be misleading, because dessert wines aren't always meant to be drunk after a meal. At times, they can be served as an excellent aperitif before a meal, drunk all by themselves on a rainy afternoon, or enjoyed during a quiet moment late in the evening. For example, you can pair an intensely sweet Sauterne or Ice Wine with *foie gras* as an appetizer. It's a tremendous way to start a meal! For our purposes, we'll stay with the term *dessert* and understand these wines are normally drunk after a meal, with dessert or by themselves.

The three styles of this wine are 1) late harvest; 2) natural intervention; and, 3) manipulated. The first style encompasses dessert wines that are made from healthy, ripe grapes. They are naturally high in sugars that are not completely fermented out of the wine. For a number of reasons, the yeast converts only some of the sugar, leaving a sweeter result. I call the second style of dessert wine "natural intervention" because the grape or the juice has been changed by a natural phenomenon so that it contains a higher proportion of sugar than normal healthy grapes. Higher sugar levels can also come from human techniques; this artificial manipulation is the third style of dessert wines.

LATE HARVEST

As grapes ripen, like any other fruit, their sugar levels rise and their acid level drops. If the grapes are left on the vines until later in the harvest season, they become very sweet. They're actually so sweet that the accumulated sugars represent too much of a task for yeast. The result is that the yeast cells are exhausted by the fermentation and quit before all the sugar is turned into alcohol. The final wine is a delightful accompaniment to many desserts like fruit flans and various cheeses.

(Note: Another way to achieve the same effect is simply to cool the wine down while it is fermenting. This stops the yeast from doing its job of converting sugar to alcohol. Once the fermentation stops, the wine is racked and filtered and the final product is a sweet, low-alcohol wine.)

The original, and finest, examples of this type of dessert wine come from Germany — especially from the Riesling grape. The sugars of the Riesling, to my palate, seem to have an indescribable richness and at the same time pleasing freshness. This grape also tends to maintain higher acid levels even when extremely ripe (sometimes even when overripe!). Riesling wine will give you a wallop of fruit and sweetness as it hits your tongue, and because of its acidity, it will seem to tingle in your mouth. Then, despite all the sugars, it finishes dry in your throat! It's an amazing taste experience.

NATURAL INTERVENTION

I could not think of another term that more adequately describes this style of dessert wines than *natural intervention*. These wines are a result of natural processes that occur in the vineyard late in the season.

The most common two ways that this sort of dessert wine is made are both results of external natural influences on the grape. The first occurs when grapes become infected by a certain fungus. In other words, they become diseased but in a good way! The word used to describe this phenomenon is *botrytized* (a short form of the fungus name *Botrytis cinerea*), also known as the *noble rot*. Yes, a rotting grape is actually beneficial to the wine. As the grapes becomes moldy, they shrivel up, losing their moisture but none of their sugars. When they are harvested and pressed, the resulting grape juice is intensely sweet, which in turn yields a lovely dessert wine. The best example of this style is Sauternes from France's Bordeaux region. However, fine examples can be found from all over the world.

Ice wine, the second example of nature creating a dessert wine, involves a fascinating approach to the grape harvest. In fact, don't even bother harvesting! Simply leave the grapes on the vines well into the winter season. Let them turn into little marbles, frozen as hard as rocks; then bring them to the winery, and watch a syrupy juice come out of the wine press. Most of the water in the grapes *stays* frozen, but the grape juice that does emerge, as thick as corn syrup, is laden with sugar. The juice is fermented[7], and the end result is delicious ice wine. My wife Sue refers to it as a "heavenly elixir." In fact, ice wine can be a dessert all by itself. Good examples are full of apricot and peach flavours, and they seem to just melt into your tongue. Ice wines originally came from Germany and were probably a result of an accident. You can imagine that one year many years ago, a vineyard owner was unable to harvest his or her grapes before the onset of winter but decided to press out the frozen grapes so as not to waste the fruit. What a surprise it must have been when the winemaker tasted the juice from the almost-forgotten

---

7   *Fermenting ice wine:* All the concentrated sugars in the juice present a problem for the yeast. Too much sugar *inhibits* yeast from even starting to do its job, so special yeasts are used to make ice wine.

crop! Today, the finest examples of ice wine come from cold climate countries, with Canada leading the way.

MANIPULATED (ARTIFICIALLY BY MAN)

In many parts of the world, winemakers add sugar to grape juice, often to achieve an adequate level of alcohol in the finished wine. The practice of adding sugar, which is called *chaptilization,* is generally used when there has been a poor growing season. It is legal in parts of France, but it's prohibited for making table wines in many other countries and regions (Italy, California, and Australia, to name a few). Sugar can also be added before or after fermentation to give a wine enough sugar to make a sweeter end product. Technically, the result should be the same as any other sweet wine with residual sugar and sweet flavours, but the added sugar doesn't match the taste profile of natural grape sugars. In general, this sort of dessert wine is inferior. The practice is most commonly used for making wine spritzers or cocktails. These wines are normally bottled in 12-ounce sizes and usually involve the addition of fruit flavours. One of my favourite uses of a sweet wine is in making what I'll call a purist's wine spritzer: On a hot afternoon, add sparkling water to the wine, garnish with a lemon or lime, and sip away!

There are many ways to manipulate a wine into appearing to be a dessert wine. (I've even heard of evaporating the water in grape juice to concentrate the sugar content.) I'll just mention one final technique that is also used to make some great table wines: The method is known as *appassimento* — an Italian term that literally means "drying out" or "shriveling." The grapes are exposed to air, sometimes laid out in the sun, so the grapes come close to becoming raisins. As you can imagine, the drying process makes the grape much sweeter. This technique can create a concentrated, fruity, high-alcohol table wine. A fine example is Amarone, a table wine made in northern Italy's Valpolicella region (it is not a dessert wine).

The drying of grapes is a technique that dates back to pre-Roman times. It was probably originally used to make raisins. Besides making grapes much sweeter before pressing, this process will preserve the fruit by removing the water content, thereby preventing spoilage. In ancient times, without refrigeration, this was very important. At some point, someone determined that these raisins could also produce a nice dessert wine. In the Côtes du Juras region in eastern France, people make *vin de paille,* or *straw wine.* The grapes are laid out on straw mats, indoors, for three months. The water inside the grape evaporates[8], which leaves a sweeter juice that renders a dessert wine. I'm not sure if it's the straw mats or the types of grapes used but this wine is one of my favourites.

Styles similar to this famous dessert wine are made all around the Mediterranean. Many have been made for centuries, and they often have fascinating stories attached to their histories.

(iii) Sparkling Wine

Champagne! What other word in the world of wine can evoke even half of the romance and mythology of this great wine? It is steeped in history and used virtually everywhere as a beverage for celebration. A friend of mine often says, "You don't need a reason for opening a bottle of Champagne, because just opening a bottle is enough reason to celebrate!" The making of this great wine is a complex and time-consuming process.

First and foremost, you must understand that all Champagnes are of the sparkling type, but *not* all sparkling wines are Champagnes. Only those sparkling wines made in the Champagne region of France can use that name. Other wines can be produced using the same method[9], but the producers should not call the wine *Champagne.*

TRADITIONAL METHOD

There are many claims to the actual discovery of Champagne. Truth be told, it could have happened in any bottled wine that still had sugar and yeast present. As the wine aged in the cellar, the yeast would continue to work

....

8   *Wine and water:* At this point, I'm sure you see a common denominator in the making of dessert wines. Whether through noble rot, frozen grapes, or drying methods, the end result is less water in the fruit and more-concentrated sugars.

9   *Champagne claims:* The three phrases used to say a winery is using the same technique as Champagne are "traditional method," "Champagne method," and "fermented in bottle."

with the sugar to produce alcohol *and* CO2. During normal fermentation, the CO2 is allowed to escape into the air. However, if the bottle is closed, then the CO2 stays in the wine, which becomes a sparkling wine!

Around the 16th century in France, people began to better understand this phenomenon, and an exact method of producing Champagne was eventually developed. The monk Dom Pérignon has historically been associated with the origins of Champagne, but that whole story is very much in dispute — we'll leave it to the historians (Pérignon is more famous for blending wines and introducing corks). What is worth preserving is the famous description of his first response to this wine: "Come quickly! I am drinking the stars." The lovely sensation of tiny bubbles married into a well-made white wine is the central part of the attraction and romance of this type of wine.

The process for making sparkling wine via the traditional method begins simply enough. Grapes are harvested and made into a still white table wine. The wine is bottled, and a little new yeast is added along with some unfermented grape juice and then enclosed (usually with a bottle cap). The bottles are now stored for anywhere from one to five years as the wine ferments again. The wine becomes a sparkling wine as the CO2 is produced and captured in the bottle. This secondary fermentation also creates sediment (lees), which stay in contact with the sparkling wine and adds some nice complexity.

At some point, the lees or the sediment in the bottle needs to be removed to clarify the wine. The traditional way to get rid of the leftover particles in the bottle involves slowly turning the bottle upside down. Each day, the bottle is tilted from horizontal to a slightly more vertical position with a gentle shaking action known as *riddling*[10] (*remuage* in French) (see Fig. 8). Once the bottle is completely upside down, the neck of the bottle is frozen by passing it through an ice cold bath. Finally, the bottle cap is removed and *degorged* as the frozen sediment pops out. A bit of still white wine and

....

10   *Mechanized riddling:* To save costs, the process of shaking and slowly tilting the wine is now done by machine, except in the traditional wineries where it is still done by hand.

sometimes extra sweetness[11] is added to replace the degorged sediment so that the bottle is again full. All that the sparkling wine needs now is a solid cork and proper packing with a wire to secure the cork. Most Champagnes are then covered by a nice foil wrapping around the top and neck of the bottle.

Fig. 8

Packaging and bottles used for Champagne are unique. For example, people sometimes wonder why the bottle requires a wire around the cork. The wire helps stabilize the cork, which would otherwise be pushed out by the pressure of the CO_2 in the wine. The early history of Champagne includes many stories of not only popping corks but also *exploding* bottles! The pressure in a bottle of Champagne can be very high, so eventually a special bottle was developed to use for these sparkling wines. You'll also notice an indentation, called a *punt,* in the bottom of Champagne bottles. It has a positive effect on the structure and strength of glass bottles. When combined with thicker glass bottles, Champagne is now safe for storage. Look for interesting large-format bottles when you want to celebrate a special occasion. A *magnum* is a double bottle size, and a *jeroboam* holds the equivalent of four

----

11 *Dosage:* To sweeten some champagne, a still wine is mixed with pure cane sugar and added after the disgorging.

bottles — the trick is getting someone strong enough to pour it! (There are even larger sizes, which are named after biblical characters).

CHARMAT PROCESS

As you can well imagine, as a result of the input (labour) costs of Champagne, the retail price is high. No wonder people keep the "real" Champagne for special events! At $50 per bottle or more, it is not an everyday beverage.

It took an Italian wine expert by the name of Federico Martinotti to devise a more cost-effective way to make sparkling wines. His process was later patented by Eugene Charmat in France. Therefore, this method is often called an Italian-French name, *metodo Charmat,* or sometimes simply the *closed tank process.*

Similar to the traditional Champagne method, the Charmat process begins with a still white wine. Instead of bottling the wine, however, the winemaker places it in a large, *closed* stainless steel tank. Then a secondary fermentation is initiated in the wine by adding sugar and yeast, and the CO_2 is trapped in the wine and unable to escape. The wine is then kept pressurized and put into bottles.[12] Cost isn't the only advantage of the Charmat process; it can be done much more quickly and often maintains a fruitier taste, though it can lack in complexity.

In most cases, the bubbles from the closed-tank process are bigger than those in a well-made sparkling wine that was created using the traditional method. The years the CO_2 spends mixing with the wine in a bottle in the traditional method will allow the bubbles (CO_2) to integrate more completely into the wine. The CO_2 actually *marries* into the wine. As an experiment, try purchasing a true Champagne and a cheap "sparkler," as I call them. You'll notice a distinct difference in the size of the bubbles when you pour the wines. The smaller bubbles in the Champagne will still be rising

12   *Other methods:* There are two other methods used to make sparkling wine. One simply involves injecting gas, as is done with pop. The other way, called the *transfer method,* is identical to the traditional method except that the wine is transferred to a vat from the bottles after the secondary fermentation, filtered, and put into new bottles.

in your glass thirty minutes later. In the cheaper sparkling wine, almost all the bubbles will have vanished within a few minutes, completely dissipated, leaving only a still white.

A NOTE ON COLOUR

The grapes used in the iconic Champagne of France are mostly limited to Chardonnay, Pinot Noir, and Pinot Meunier. The latter two are actually red grapes! (Remember, almost all grape juice is white when pressed quickly.) All over the world, these same grapes are usually used to make quality sparkling wine. If only Chardonnay is used, then the label may say "blanc de blanc," or "white from white," which simply denotes that the winemaker used only this white grape. Conversely, when only a red grape is used, the wine label often reads "blanc de noir" or "white from dark" — these wines are also a favourite of mine. They're usually made with only Pinot Noir grapes (but just like other Champagnes, they are still white in colour).

Some sparkling red wines exist, although they're relatively rare. The process used to make them is the same as those used for whites. However, I do not know of many red sparkling wines made by the traditional method. The best sparkling reds I've tasted come from Australia and Burgundy, France. They are quite interesting and make a surprisingly good match for green salads with fruit garnish.

When choosing a good bottle of sparkling, I would always recommend those made in the true Champagne method. However, somewhat controversially, I'd say that the best *value* doesn't always come from France. I am particularly fond of sparkling whites made in California and Washington State — especially those using the traditional method. Look to many cooler regions that have conditions similar to Champagne, France, to find other fine examples (Northern Spain, New Zealand, Niagara region in Canada, and New York's Finger Lakes area). Regardless of your choice, there is no better way to begin a meal!

VINTAGES AND BLENDING CHAMPAGNE

In general, the word *vintage* simply refers to the date of the harvest. In Champagne, France, there are some years in which the harvest is deemed to be inferior. In those years, most producers will not "declare" a vintage. This means that the wine they make that year will not have a year on the bottle's label. Instead, that wine will be put into a blended wine usually made with various years' products. In fact, the vast majority of Champagne is used for these non-vintage (NV) wines. Each winery, or Champagne House, will always blend various years to make a standard wine. This house style is meant to be the same year after year. The winemaker, or *chef du cave,* will sometimes blend more than 30 wines from various vineyards and different years to create the flagship house style.

In terms of quality, you can expect that vintage Champagne would be better than NV Champagne because the grapes themselves are higher quality. In a good year, the grapes have a perfect blend between the sugars and acids. The winemaker will extract the right amount of fruit during fermentation and then add complexity in aging the wine on the lees during the secondary bottle fermentation, giving the final product just the right amount of flavours and bracing acidity to give it a fresh, lively sensation that is amplified by the bubbly effervescence. Vintage Champagnes, with the year clearly marked on the label, will have considerable aging potential as well. A NV Champagne will improve with some aging, becoming softer and richer in taste like Vintage Champagne, but NV aging potential is usually limited to five years from its release date. After this initial aging period, the wine will become less vibrant and eventually rather bland. My best advice is to keep the Vintage Champagne for special occasions and drink NV Champagne and the other sparkling wines young — and often!

(iv) Fortified Wines

The word *fortify* means "to make strong" or "to reinforce." In the world of wine, it simply means adding alcohol to a wine. The result of fortifying wine is an increased alcohol level that usually reaches a total of between 16% and 20% alcohol (compared to the 12% average, common for most wines). A wine that has this level of alcohol is *stabilized* and will not deteriorate the way table wines would. Wine, like people, grows fragile as it ages. Adding

extra alcohol to wine is akin to a person finding the fountain of youth[13] —
these wines will age for a very long time, some even more than a hundred
years. The four major styles of fortified wine are Port, Sherry, Madeira,
and Vermouth.

PORT

This fortified wine comes from Portugal, where it is called *Vinho do Porto*. It
is almost always red and very sweet, and it's usually used as a dessert wine
or post-dinner sipper. Port is also made in many other countries, and those
wines should actually be labeled "Port-style." Unfortunately, many places
simply call it *Port*, which is misleading. True Port is made only in Portugal's
Douro River region, where the city of Oporto is located (just as real
Champagne must only come from the Champagne area in France). I highly
recommend a visit to this region. The Douro River Valley is an endless sea
of steep terraced vineyards, and the views are spectacular.

The process involved in making Port is quite interesting compared to
that of other fortified wines. Instead of fermenting the original wine base
until all the sugar is turned into alcohol by yeasts, Port producers stop the
fermentation when the wine reaches only 4% alcohol by adding alcohol,
usually a neutral tasting grape brandy. The alcohol stops the yeast from
doing its work of converting sugar to alcohol, because most yeasts die when
a wine exceeds 15% alcohol. The end result is an intensely sweet and delicious wine.

There are many styles of Port, including white ones, which are less sweet
and make a perfect aperitif. The main difference in Port styles has to do
with the aging technique. The first major style is a result of aging the wine
in wood barrels. This allows the Port to mature quickly due to greater exposure to oxygen. Just like an apple left out on the counter, the wine takes on a
brownish colour and develops a nutty, complex flavour.

....

13   *Spirits and life:* The analogy to finding a youth tonic is not as farfetched as you
     might think. The original spirits were referred to as *eau de vie,* or "water of life."
     Many of today's words for distilled spirits have their etymological roots in the
     word for *water* or the phrase *water of life* (e.g., vodka, whisky, aqua vitae).

The other main style of Port is aged in a bottle. This category contains many substyles, but the most interesting are the Vintage Ports. As with Champagne, some year's grape harvests are declared top quality, so a winery decides to make the Port solely from that crop instead of blending it with other vintages. The wine is then made as usual and aged up to two and a half years in barrels before bottling. A Vintage Port will age in the bottle for many years. In fact, such wines are not even deemed ready (i.e., having developed full character) until ten years or more. I've enjoyed many Vintage Ports that had aged more than thirty years, and quite astonishingly, they lose a bit of sweetness while still tasting fruity and very mellow on the palate. There are tales of the discovery of 19th century Vintage Ports that when opened still showed beautiful character and overall health.

SHERRY

Traditionally, Sherry comes from the south of Spain. The Spanish name for this fortified wine is *Jerez* (pronounced hair-*eth*), which is the name of the place where it is made. Sherry-style wine, like Port, is also made around the world and, again, such wines are incorrectly called "Sherry."

Like Port, there are many substyles of Sherry. They can be bone dry (no sugars left in the wine) or extremely sweet. The grapes used in sherry are whites, with the most prominent one being the Palomino. The juice of the grapes is fermented in much the same way as normal white wines. Then the wine is lightly fortified with alcohol and allowed to age for a year on average. Finally, the wine is put into a *solera,* which consists of rows of open wooden barrels stacked one upon the other. There can be 15 barrels in each row, and the rows can be stacked 4 or more high. The new wine goes into the top barrels. As the wine ages, it is transferred to the row of barrels below, and so on. When the wine reaches the lowest row, it is ready for bottling. The wine therefore is a blend of many years. It is also continuously exposed to air as it ages; this constant oxidization is one of the main factors — along with the grape and yeasts — that gives Sherry its nutty, complex flavour. This characteristic develops in the solera.

Fino Sherry: This substyle of Sherry is a result of a phenomenon that originally caused Sherry makers great concern. A wild yeast that thrives in this part of Spain frequently spontaneously blooms on the surface of the Sherry sitting in the open barrels of the solera. This layer of yeast is called *flor*, which translates as "flower" in English. At first, winemakers thought the wine had gone bad, but as the wine sat under this yellowish covering, it created light, dry, and refreshing wines. This flor yeast is the distinguishing characteristic of the Fino Sherry styles. These Sherries are generally more delicate-tasting wines made in a bold, very dry style, and they should be drunk within a few days of opening the bottle.

Olorosa Sherry: This style of Sherry is made without the "infecting" flor yeast. It, too, is made in a solera and will be quite old when it is finally drawn off the bottom barrels. Oloroso Sherries tend to be more aromatic and, except in the rarest cases, are sweetened at the time of bottling by adding sterile grape juice. They are also fortified to a greater degree (i.e., more alcohol is added, resulting in 18–20% alcohol levels). Due to the higher sugar and alcohol content, you can drink Oloroso-style Sherries (including Cream Sherries) months after opening a bottle. Although I don't recommend it, people often store these Sherries in fancy decanters.

MADEIRA

"Have some Madeira, m'dear" is a famous old song lyric associated with the seduction of women — not a bad thing if all parties are onside and consenting! Madeira is from — you guessed it — Madeira, a collection of spectacular Portuguese islands off the western coast of Africa. Similar to Sherry and Port, this fortified wine can be dry and used as an aperitif, or it can be made in the more common sweet style, which is then used as a dessert wine. The unique step in making Madeira is that of heating the wine. The reason for the heating is related to the long history associated with wines from these beautiful islands (often called "the garden islands" due to the profusion of flowers).

Madeira has been a center for trade and shipping since the 1500s. Among the many other goods, local wine was often included in the cargo. However, since Madeira is near the equator, the wine would often spoil during its transport. Adding alcohol — that is, fortifying the wine — solved the spoilage issue. The hot equatorial sun heated the fortified wine, changing the taste of the wine but in a good way. The result was a creamy caramel taste with nutty undertones and a refreshing, bracing acidity.

Today the wine is purposely heated[14] and exposed to air. This oxidization further stabilizes the wine and produces its signature flavour. In fact, when regular table wines become oxidized due to poor storage or defective corks, the wine is said to be *maderized.*

Madeira is made into four major substyles, each named after the grape used to make it. They go from the sweetest, Malvasia or Malmsey, to Bual and Verdelho, and then to the driest, Sercial, which is fermented to the point of being almost completely dry. The sweeter versions are made sweet by simply halting the fermentation so that some natural grape sugars remain (similar to making Port). Then a neutral grape spirit, technically a brandy[15], is added to fortify the wine. Madeiras have been known to age well over 200 years! Yes, a little Madeira for me, my dear. (Also one of my favourite places on earth!)

VERMOUTH

In the ancient world, it was common practice to flavour a wine with various concoctions, including herbs and spices. I have conducted wine tastings where I've used ancient Roman recipes to explain some of the historical background of Vermouth, and it is fascinating to see people's reaction to the flavours. This practice was widespread in earlier times due to the quality of the wines. Some wine tasted awful due to poor water, poor storage, or just plain lousy winemaking, so various ingredients were added to make the wine palatable. Over time, a style of wine that was the forerunner of Vermouth developed. Such wines were often used as a medicinal beverage because of the beneficial properties associated with the herbs and spices.

....

14   *Cooking the wine:* Madeira is brought up to temperatures of 140°F.

15   *Distilled spirits:* All distilled products come from a previously fermented beverage. The alcohol is then concentrated, or *distilled,* by evaporating the water. If the fermented beverage is wine, then the distilled after-product is brandy; whisky is made from beer.

The origin of the name *Vermouth* comes from one of its historical ingredients, wormwood. The German word for this ingredient, *Wermert,* gave birth to our word *Vermouth*. In ancient times, wormwood was used for treating stomach disorders. Over the centuries, all kinds of other medicinal ingredients were used in wines to create various tonics. Sometimes these wines were fortified for additional medicinal effect. Eventually, producers created their own, well-guarded procedures and recipes for their proprietary brands. In France, Vermouth is generally white and dry and is made with roots and herb mixes. Italy, the other major producer, tends to make sweeter red Vermouths that also include various ingredients.

Vermouth is a great addition to many food recipes and is also used in an assortment of cocktails. Dry Vermouth with a lemon or lime garnish is an excellent aperitif prior to a hearty meal.

(v) Wine Classification Table

Types	Styles
White /Table	Plain, light, neutral whites ("winey")
	Fruity and flowery (aperitif wines)
	Complex and full-bodied (pronounced character, food and cheese wines)
Rosé /Table	Dry (French)
	Sweet and fruity (Portuguese, Californian)
Red /Table	Plain, simple bulk (quaffing wines)
	Young, fruity smooth (aperitif and food wines)
	Mature, complex, balanced (the epitome of great wine!)
	Concentrated, powerful reds (dinner, game meats, and strong cheese wines)
Dessert Wines	Late harvest
	Natural intervention
	Manipulated processes (artificial)
Sparkling	Bottle fermented/traditional method (dry or *brut*, semi-dry, and sweet styles)
	Closed tank/Charmat process
Fortified Wines	Aperitifs (dry Sherry and Vermouth — often used in cocktails)
	Dessert (Port, Madeira, sweet Sherry)

III. THE HISTORY OF WINE

*"Wine was created from the beginning to make people joyful, and
not to make them drunk."* —Ecclesiasticus (Book of Sirach)

Wine is steeped in myth and has a fascinating history. Its discovery, which was probably an accident, would have occurred at the time humans first gathered grapes. As mentioned earlier, we'd need only an absentminded caveman or woman to forget about the bunches of grapes in the corner — soon the grapes would have rotted, leaking juices that would turn to wine as a result of coming in contact with the wild yeasts on the grape skins. A rather prosaic beginning for our marvelous beverage!

The more interesting history is portrayed in the legends and myths associated with wine. Some of these stories start with ancient Egyptian paintings and statues celebrating the virtues of wine. Frequently, they show wine being offered to the gods. In other paintings, wine is depicted as being controlled and available only to the elites and the priests (the poor peasants had to drink beer). Various other drawings show the details of wine production and its use in daily life. From the earliest times, people considered wine beneficial to one's health. The Egyptians carefully documented winemaking techniques. There are even records that show jars of wine were labeled with the year, vineyard, owner, and winemaker.

Wine has always been closely associated with the divine. It was considered a magical beverage. Drink enough of the stuff, and you can communicate with the gods. Egyptian priests would enter this holy state on behalf of their brethren. I bet a lot of folks wanted to sign up for that job!

The Greeks also incorporated wine into some of their rituals. One of their gods, Dionysius, was known as the God of Wine. Myth has him born of a god and a mortal, eventually killed, and then reborn — echoing aspects of the Christian story. Philosophers participated in *symposiums* (meaning "drinking together") where the wine would elevate their conversation for a while, and after a little too much wine, these get-togethers ended with a bunch of off-colour poems. The Romans turned Dionysius into Bacchus, a less dignified god but a boozer nevertheless. Over time, this association of wine with the divine continued into the development of Christianity.

Let us leave the myths and the realm of the gods and turn to the major developments in the history of wine. Many terrific historic books about wine are available if you're interested in delving further into this fascinating subject. I highly recommend Hugh Johnson's *Vintage: The Story of Wine*.[16] It is a great read, very entertaining. For our purposes, I'll give you the Reader's Digest version of history and outline what I consider the seven main developments in the multifaceted past of wine, starting with the Romans.

(i) The Roman Empire

The Egyptians and the Greeks were both deeply immersed in wine production. As we've seen, they incorporated wines into many aspects of their culture. Soon a new country, now known as Italy, started to embrace the grape. In fact, the Greeks called Italy *Oenotria*, meaning "the land of vines." As the Roman civilization became more sophisticated, so did its

16 *Story of Wine:* The video of this book — titled *Vintage: A History of Wine* — is also terrific. I used some of the episodes for a university course I taught in Europe in 1990.

wine industry.[17] I will concentrate on two of their major contributions: the spread of wine knowledge and the expansion of its popularity.

WINE KNOW-HOW

Earlier civilizations knew the basics of making wine, but the Romans were the ones who refined that knowledge and recorded it in a more detailed manner. Many of their written techniques were followed well into the 19th century with few, if any, changes! Thanks to sites like Pompeii[18] and writings from the first century, we know a lot about what the Romans were doing with wine, including the wine trade with Rome's neighbours.

In the earliest days of the Roman Empire, Greek wine was considered superior and was much sought after among the Romans. Soon, however, the Romans surpassed the Greek winemakers. Even now, Italy is a wine powerhouse, while Greek wines are of very little significance in the modern world. The Romans were methodical and passionate about their wine — and they still are!

The practice of winemaking moved from art to science in the Roman world. The Romans identified the best grapes and soils. They rated the best vineyards and vintages. They carefully developed and recorded wine practices as early as the 2nd century BC. An agronomist named Columella wrote an influential major treatise on grape-growing in the 1st century AD. By this time, the Romans had become masters of wine, and soon they were going to spread not only that knowledge but also a genuine appreciation of wine throughout Europe.

EXPANDING THE LOVE OF WINE

The people around the Mediterranean didn't just enjoy drinking wine — they really *valued* it. For many years, wine had been an important object for

....

17 *Wine industry*: To say the Romans had a wine "industry" is not a misplaced description. Wine was bought and sold in great volumes throughout the Empire.

18 *Volcanic site of Pompeii*: This city was suddenly covered by ash from a violent eruption of Mount Vesuvius in the year AD 79, and the ruins were beautifully preserved.

trade. The Romans took their wine to the lands they conquered. At first, the wine was probably only for the marching legions. The wines were originally transported in large clay pots known as *amphorae*[19]. These large containers were excellent for storing wine, and they accompanied the Romans to every part of the known world (see Fig. 9). Soon they introduced wines to the conquered tribes whenever they traveled. Of course, these people embraced this delightful beverage (compared to the crude beers they were used to drinking) and were willing to pay for it. Before long, the wine trade was thriving in the areas we now know as Spain, Portugal, France, Germany, and even Britain.

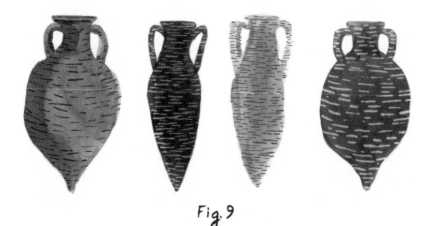

Fig. 9

At some point during the peak of the Roman Empire, grape vines were planted in the conquered regions. The logic was simple — why transport wine if you can grow grapes and make wine where you're stationed? Soon the wine from the far-flung regions of the Empire became quite good. In fact, these wines had different and sometimes better characteristics than Greek and Roman wines — maybe fruitier or perhaps with a refreshing acidity — so eventually these wines were being sent back to Rome! We know this because, among other archeological sites, the ruins of Pompeii yielded amphorae of wine from many other regions in northern Europe.

---

19 *Transporting wine:* Around the same time the Roman wine trade was flourishing, wood barrels were also beginning to be used to ship wines. They were sturdier than the amphorae, which were susceptible to breakage.

One of the main outcomes of Rome's influence in the ancient world was the establishment of a wine culture in the countries north and west of Italy. The Romans stuck mainly to river or ocean routes for trading purposes. Floating shipments of goods was easier than moving things overland. Today you can follow those ancient trade routes and find or stumble upon the greatest wine regions in the world. Just think of the areas planted in vines by those intrepid conquerors! France's Rhône and Loire valleys, Bordeaux, Burgundy, Alsace, and Champagne; Spain's Sherry and Rioja regions; the Douro River Valley in Portugal; Germany's Mosel and Rhine valleys . . . the list goes on and on. Planting vineyards and spreading a love of wine was surely one of Rome's greatest cultural contributions.

(ii) Christianity

You can find over 500 references to wine in the Bible. The first miracle Jesus performed was turning water into wine. The Old Testament mentions that Moses (approximately 1400 BC) sent out two men to explore the Promised Land, and they returned with a cluster of grapes that they carried on a pole between them (Numbers 13:23). Proverbs 31:4–7 contains my favorite quotation: "[. . .] give wine to he who is perishing, in bitter distress, and let them forget their poverty and remember their misery no more." Sounds prudent!

The critical point in Christianity's history with respect to wine came when the drinking of wine became a Church sacrament. Known as the *Eucharist*, this rite has its origins in the story of the Last Supper, when Jesus broke the bread and said, "This is my body"; sipped the wine, saying, "This is my blood"; and asked his followers to partake in these activities regularly in remembrance of him.[20] From that point on, wine accompanied the path of Christianity's growth throughout the world.

Regardless of the origins of this sacred rite, the Church needed wine for its services, and it fell upon the priests in many parts of Christian lands to procure wine. Eventually, the priests themselves made the wine. They would first pick a location suitable for vineyards and *then* build a church or abbey in the middle of the vines. The Emperor Charlemagne in the 8th century bestowed a diploma upon a courtier with the following inscription: "He has built churches and planted vines." In the 14th century, Pope Clement V moved the papacy to France. Of course, vines were planted around this "new home of the pope" (in French, the *Châteauneuf-du-Pape*).

Beginning in the 15th century, Christian Europe went off to discover the "New World." Along with colonization and evangelization came the planting of vines and the making of wine. Portuguese missionaries introduced wine to Brazil and Japan. Spaniards took it to North and South America. The French brought wine to Canada and Africa's Cape of Good Hope. Even Russian domination of parts of the East allowed the Orthodox Church to raise wine consciousness in the "dry" Islamic countries.[21]

MONASTERIES AND THE CATHOLIC CHURCH

The Catholic Church has often been described as a great force in the spread of viticulture around the world. Priests and monks preserved and perfected the art of winemaking throughout the Dark and Middle Ages. An example of this phenomenon occurred in Burgundy. Monks, the Cistercians, became expert winemakers — and drinkers — beginning in the 12th century. They

20 *Wine and blood:* Scholars have debated the true origin of this sacred rite of Christianity, which equated wine with blood. To some, it echoed cannibalism. We do know that Greeks and Romans had started using wine sacrificially instead of blood, which was common in earlier times (including stories in the Old Testament). The Bible even refers to wine as the "blood of the grape."

21 *Islam:* The use of alcohol was forbidden in the Muslim world two to three centuries after the founding of the religion. However, there have been some imbibers over the centuries — perhaps most notably, Omar Khayyám.

aggressively expanded in the region, and each time, they planted more vines. The monks diligently recorded all aspects of winemaking and classified the best areas. Stories tell of the monks actually *tasting* the dirt to determine whether the soil would be good for grapes! In very little time, their wines became well-known for their richness and finesse.

Eventually, wine became necessary not only for the Eucharist but also as a source of funds to offset the costs of housing monks and the many followers of the religion. Besides requiring more wine as the Church grew — no wine, no Mass — quality was increasingly important for the buyers. This greater emphasis on quality, in turn, created more demand and higher prices. Too bad the monks couldn't take advantage of large harvests and inventory the wine for years to come as valuable assets for the monastery. Unfortunately, wine would spoil when stored for too long because of exposure to the air. It took a clever monk to solve that problem.

(iii) Corks and the Glass Bottle: The Perfect Marriage

In ancient times, wine was stored in earthenware vessels called *amphorae*. Trouble was, these containers allowed oxygen to reach the wine, which would lead to early spoilage. A number of techniques were employed to (a) stop the spoilage (this included heating the wine and methods to stop air from reaching the wine) or (b) disguise the off-flavours (this strategy involved adding various herbs and spices to make the oxidized wine palatable). In Greece, the winemakers would often coat the inside of the amphorae with resin to minimize the porous nature of the clay vessels and then use a cloth soaked in the resin stuffed into the opening to act as a stopper[22]. Stories tell of a Roman wine made in the time of the consul Optimus that was said to be just fine to drink 100 years later. If the story is true, the wine was undoubtedly stored an impermeable container.

GLASS BOTTLES AND WINE

Glass makes a perfect storage vessel for wine. It is inert, thereby giving no flavours to the wine. It is also completely impermeable, preventing any oxidization. Wine bottles are also easy to store when laid down in stacks. Glass bottles were used since the time of the Phoenicians; however, these containers were used only to *serve* wine, not store it, because of one big problem: There was no reliable stopper.

CORKS

Although cork was used from the time of the Egyptians, it was not used as a stopper for wine storage containers. The Greeks and Romans used it for shoes, for ship construction, and for fishing buoys, among other purposes. It wasn't until the 1600s that the French monk Dom Pérignon started using it with glass bottles. Pérignon was mythically credited with discovering Champagne, but his real reputation was as a master blender of wines that created better and more consistent wines. The cork also helped him to keep the lid on the gassy sparkling wines, as other stoppers would regularly pop out. The cork was more dependable. Regardless of Pérignon's other supposed contributions, from that period onward, the cork became the standard closure of wine. It was a marriage made in heaven.

Cork properties and sources: Cork is an unusual type of wood. The magnificent thing about cork is its ability to expand when wet, stay healthy, and not rot for many years (it is also a soft wood, which allows for easy extraction from bottles!). As a stopper for wines, cork prevents all but the smallest amounts of air from entering the bottle. When the wine bottle is laid down horizontally, the cork expands firmly and continually touches the sides of the bottle's neck. Combine this substance as a stopper with a secondary

22 *Retsina:* A wine still made in Greece adds pine resin to mimic the taste that resulted from this practice.

enclosure like wax, metal foil, or plastic, and the wine is encapsulated in a safe, almost inert, container.

So where does cork come from? It is the bark of a particular type of oak tree called *Quercus suber*, which is a slow-growing evergreen. The spongy outer bark is the part of the tree harvested for corks. The oak must be at least 20 to 25 years old before its bark is removed for the first time (this harvesting continues every 10 years, until the tree dies at around 200 years old!). Once harvested, the sheets of the bark are stacked to dry out and then are boiled in a fungicide to sterilize the wood. After more storage, the wine corks are stamped out like plugs — and then finally, the corks are graded and shipped to wineries.

Today over 50% of corks come from Portuguese forests. Cork production is a sustainable industry. It preserves valuable habitat for many animals and, when compared to all ancient and modern stoppers, cork is the most environmentally friendly substance.

Aging revelation: This wonderful juxtaposition of glass and cork resulted in a new potential for aging wines carefully. Whereas wine stored in wooden barrels would soften and mature over time, the result was often unpredictable. The wood could add too much flavour at times. Other problems included rapid maturation, sometimes leading to an outright spoilage due to oxidization, and there would often be a 5% *yearly* evaporation of the wine — not only a pity but also costly.

A properly enclosed bottle protects the wine from oxygen. The various elements in the wine are allowed to interact slowly over time. They begin to combine (polymerize) into new compounds, flavours, and smells never experienced before in young wines. The gushing descriptions from today's wine experts are a result of aging: "It smelled of violets, with hints of licorice and tobacco, a touch of barnyard, marvelously integrated tannic structure, with bracing mouthfeel, and then more floral-tinged dark berries at the back of the mouth and a beautifully balanced, complex lingering aftertaste . . . a finish that lasted into the following morning." Don't you just love it?

Screw Caps: A recent phenomenon in bottling wine has been the introduction of metal screw caps to replace corks. Some people argue that screw caps will prevent the beneficial slow oxidization that corks facilitate, as corks are minutely porous, which in turn allows for the aging process at a timely rate, leading to some exquisite wines. However, the big problem with corks is that some contain bacteria that spoil the wine (sometimes as high as 10% of the time!). Screw caps eliminate this problem. And truth be told, the small amount of trapped air in the neck of the bottle (known as the *ullage*) seems to be sufficient to allow the wine to slowly oxidize and age safely. The big advantage, of course, is the ease of opening the bottle itself. I must admit that when we are having a dinner party and the wine is flowing, I am very pleased when the next bottle to be opened is a screw cap. It's so convenient! No doubt the cork versus screw cap debate will rage on for years to come.

(iv) Phylloxera

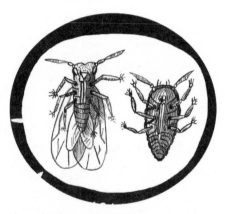

Who would have thought that a tiny little bug would give new meaning to the word *louse?* Well, just such an almost-microscopic, yellow aphid-like insect devastated the vineyards of Europe in a way that was (pardon the pun) truly lousy.

During the 19th century, there was a lively exchange of grape vine cuttings across the Atlantic Ocean.[23] European colonizers were anxious to plant grape vines that would produce wine just like those at home. Unfortunately,

---

23 *Thomas Jefferson:* The third president of the United States, author of the Declaration of Independence, was an avid wine drinker and was very knowledgeable. He repeatedly tried and failed to grow European grapes at this home, Monticello.

plants continually died after just a few years in the New World. No one was sure why at the time, but meanwhile, people were also bringing back Native American vines to plant in Europe. They wanted to see how the grapes would grow and whether they would make better wine than in the colonies. The exchange was Pandora's Box of wine being flung wide open.

The first cuttings arrived in France in the early 19th century. Around the 1860s, the majority of France's native grape wines were withering and dying. Soon the rest of Europe's vineyards were also infected, causing a crisis of unimaginable proportions. Why? Well, not only was wine a staple in most diets, but wine regions in France and Italy depended greatly on the wine trade economically. By some estimates, over 50% of people were employed in that industry in places like Tuscany and Bordeaux.

Determining the problem took years of experiments and research.[24] It turned out that the Native American vines came with a little passenger. The blight known as phylloxera was caused by a tiny louse (*Daktulosphaira vitifoliae* in Latin). Many attempts were made to kill the insect, which attacked the roots of the grape vines. Vineyards were flooded to drown the louse with some effect. Carbon bisulfate was injected into the soil to kill the louse, again with some success. But overall, phylloxera spread by way of cuttings, farm equipment, and peoples' shoes and even in the grape juice transported to neighbouring regions. Despite the best efforts of all concerned, the blight had ravaged most of the vineyards in Europe by 1900.

TWO SOLUTIONS

Many experiments in Europe were designed to eradicate phylloxera; however, even the partially successful techniques were only short-term solutions. Stop the treatment, and the louse would return. The solution had to be far more radical than simply killing this persistent insect. The trick was to build an immunity to the louse.

Hybrid grapes: The scientists of the day knew that European vines planted in the Americas eventually died. Yet planted right next to them were native

....

24   *Bounty on a louse:* The French government offered 300,000 French francs (around 1.5 million dollars today) for any scientist who could cure the blight.

grape vines growing happily and flourishing. This puzzling situation suggested a resolution: Someone realized that if they crossed the European species, *Vitis vinifera*,[25] with various American varieties (such as *Vitis labrusca* and *Vitis riparia*), then the new vine might take on the resistant attribute of the American species. Fortunately, some of these hybrids were indeed resistant to the louse. New grapes, often named after scientists, like Vidal, Seyval, Baco, and Maréchal Foch, soon became an option for wineries. Unfortunately, they were rarely of equal quality to the *Vitis vinifera* from Europe.

Grafting: For centuries, farmers knew that you could graft different plants together. A pear limb can be grafted onto a peach tree and still produce pears. This reality suggested the ultimate solution to the phylloxera infestation in Europe. Certain root stocks native to the Americas were chosen for grafting. The European cuttings were attached to those root stocks and started producing grapes just like in the old days (although some people argue that the quality is inferior). I believe that the wines of today, made from the classic grapes, are probably as good as they have ever been, and now virtually all grape vines in the world have North American roots!

MODERN-DAY WORRIES

There are still areas of the world where phylloxera did not affect the original grape vines. Most notably, Australia, Chile, parts of England, and Washington State still have ungrafted vineyards of 100% *Vitis vinifera*, the European species. Most of these unaffected regions are isolated by mountains and often have dry climates and sandy soils, both of which slow down the louse. But that aside, growers must always be on guard for an infestation, the most recent of which was the "Great Vine Plague" that arrived in England by way of Morocco in 1972. It spread throughout English vineyards, destroying as much as 60% of all vines in the country. This louse is no doubt still with us 150 years after its first arrival in Europe.

....

25 *Vitis Vinifera grapes:* All the classic grape varieties belong to this species i.e. Chardonnay, Merlot, Cabernet Sauvignon, Riesling, etc.

(v) Louis Pasteur

"Wine is the most healthful and hygienic of beverages" —Louis Pasteur, *Studies on Wine*

Although wine was safer to drink than water in earlier times, wine still had its problems. In the mid-1800s, Louis Pasteur revolutionized the techniques associated with winemaking. He began his research with questions about why beer and wine turned sour. Pasteur's studies helped identify how yeast acts on sugar to create alcohol and carbon dioxide. During the course of his research, he noticed not only round yeast cells but also rod-shaped organisms (bacteria). It was the bacteria that caused the beer to sour. Further experimentation revealed that simply by heating up the *wort,* the unfermented beer, one could correct the problem.

Soon Pasteur's technique was applied to grape juice. A light heating took place before the yeast was added, so no harmful bacteria was available to sour the wine. Louis Pasteur's approach represented a watershed in the world of wine — a time when the traditional art of making wine became even more married to the world of hard science. Pasteur's work was enormously beneficial for the wine industry. Because of his work, wineries became more stringent in their sanitation practices, eliminating harmful germs from their equipment and premises wherever possible. Pasteur saved the wineries thousands of francs, as these new practices produced wine that was of dependable and consistent quality year after year.

Heating the *must,* unfermented grape juice, is no longer a common practice. Today, many wineries add small amounts of sulfur dioxide (SO2) to kill or at least arrest harmful bacteria (and wild yeasts). The SO2 can be added before or after fermentation, depending on the type of wine and winery practices. Generally, most of the sulfur will dissipate with age.

Pasteurization today is mainly limited to the beer industry (and of course milk products). After being bottled, the beer (or sometimes wine) goes through a *flash* pasteurization, which is a quick heating in a hot water bath at the end of the bottling line. This practice is viewed upon with some dismay by purists. Factory wineries and big breweries will sacrifice freshness and quality for the sake of stabilizing the end product. Globalization has made the ability to travel more important than taste or quality — a sad but true assessment, I believe.

Pasteur gave us knowledge regarding yeasts and harmful bacteria. That is his great legacy. However, the practice of pasteurization is not needed for fine wine.

(vi) Prohibition

"It is a fight with an enemy more mighty, more merciless, more beastly, more fiendlike, more diabolical than Teuton" [26] —Reverend C.A. Williams, on the fight against alcohol, 1914

Prohibition in the late 19th and early 20th centuries was by no means a new concept. For example, in countries where Islam predominates, the

26   *Teuton:* The Teutons were an ancient, very fierce Germanic tribe of people.

consumption of alcohol has been banished for more than ten centuries. In North America, the enforcement of official, government-sanctioned Prohibition (mainly in the 1920s) was preceded by the Temperance Movement. I'll start with a brief view of the years leading up to Prohibition and then describe its sometimes lingering repercussions.

THE TEMPERANCE MOVEMENT

Tales of the drinking habits of English and American populations in the 19th century are sometimes astounding. Certain segments of society, especially male, frequently consumed prodigious amounts of alcohol. For example, American Founding Father John Adams began breakfast with a glass of hard cider. Businessmen were known to interrupt their day to observe the "elevens," an 11 a.m. round of spirits, normally whiskey — then they would stop for drinks on the way home and often end up in a tavern after the family dinner. Had statistics been kept, we probably would have found that domestic violence was widespread and that productivity in the workplace was less than stellar. In fact, one of the most successful arguments for prohibition was economic: A sober nation is a productive one.

In the U.S. and Canada, the Temperance Movement gained its greatest force when the various groups that had formed to moderate the consumption of alcohol coordinated their efforts. Famous examples of these groups include the Cleveland Women's Christian Temperance Union, which was formed in 1874, followed by the Anti-Saloon League (1895). These groups and others like them found tremendous support from various churches (specifically Protestant ones). Their main goal was to restore peace at home and promote general family health. As the movement gained momentum, the goal became absolute prohibition rather than moderation. A beneficial side effect of these women's groups is that they laid the groundwork for the movement for universal suffrage.

By the early 20th century, a number of states and Canadian provinces had gone "dry." The culmination of the Temperance Movement in North America was the enacting of laws against the sale of alcoholic beverages. In other words, *prohibition*. However, exceptions for religious ceremonies, homemade wine, and health-related uses were allowed under the laws. As you can imagine, applications to use wine for Communion and as an elixir to cure various ailments rose spectacularly when Prohibition took effect. It also unleashed a torrent of other unintended consequences.

THE PROHIBITION YEARS

In both Canada and the United States, Prohibition lasted around twelve years (mainly during the 1920s in both countries). The immediate effect of the laws was to squash a growing wine culture. The number of wineries plummeted. Output was mainly limited to sacramental wines and various elixirs or tonics for the sick (a lot of people felt "sick" in those years!). After Prohibition ended, it would take seventy years for the number of quality wineries to match those in existence before the laws were introduced. Amateur winemaking also became very common during Prohibition.

Grape quality: One unintended consequence of Prohibition was a decline in the number of vineyards planted with classic wine grape varieties. As a general rule, better-quality wine grapes do not travel well. So as home wineries increased their demand for grapes, the vineyards ripped out varieties like Chardonnay and Cabernet Sauvignon and planted hardier types. These new varieties had been used mainly for table grapes or raisins. Because the quality of the wine was inferior due to poor grapes, North Americans not only lost quality vineyard plantings but also lost their appreciation of good wine. One amusing development was that although the number of wineries decreased, the actual acreage of vineyards increased dramatically during Prohibition.

Wine consumption also increased.[27] Once again, government intrusion into public life ended up achieving the opposite of its intended goal.

Crime in the U.S.: The Roaring Twenties was a time of revelry in the U.S., exemplified by the flapper and the speakeasy. Not only was the wine industry decimated, but so was regard for authority. People flocked to speakeasies, underground bars, and anywhere they could get alcohol. Gangsters owned most of these operations, and the American mob flourished like

....

27   *Homemade wine:* The U.S. law allowed the head of the household to produce 200 gallons of wine per year!

never before. Law officers were hired by the thousands to enforce the law but seemingly to no avail. As I mentioned earlier, consumption of alcohol rose during Prohibition.

Rich Canadians: Canada took a different approach during its years of Prohibition. Wineries were allowed to continue making wine with some restrictions. More importantly, breweries and distilleries were allowed to continue making beer and spirits (mainly whiskey and rum) but only for *export*. Fortunately, they had customers with a very high demand for their product: thirsty Americans. Bootlegging across the border, by truck or boat, became increasingly common. Stories of intrigue, near escapes, spectacular smuggling busts, and deaths abounded. And many booze barons laid the foundation for family dynasties that are still predominant in Canada. Thanks to the American gangsters, a ridiculous movement, and the continued demand for alcohol, these leaders of industry became very wealthy. Not surprisingly, their partners in crime[28] are rarely mentioned in family histories.

TEMPERANCE TODAY

Temperance is still with us today. Obvious examples of prohibition persist in places like Utah, and various movements are still actively working to decrease alcohol use. Mothers Against Drunk Driving (MADD) is a group that is doing some good work to promote sensible consumption of alcohol. Carol Lightner founded MADD in 1980 in California after she lost a teenage daughter in a drunk driving accident. The group originally focused on stopping drivers from getting into their vehicles under the influence of alcohol. While we all agree that people shouldn't be driving after a few drinks, the campaign has morphed into a strident, anti-alcohol anytime movement. In 2002, Lightner stated that MADD "has become far more neo-prohibitionist than I had ever wanted or envisioned." She continued, "I didn't start MADD to deal with alcohol. I started MADD to deal with the issue of drunk driving."

....

28   *Al Capone in Canada:* I live in Guelph, Ontario, where we have a hotel, The Albion, that hosted Al Capone on some of his "business" trips — no doubt, at times, relating to buying booze.

Public opinion tends to ebb and flow regarding the positive and negative effects of alcohol. In America, the attitude often reflects the Puritan roots of its early colonization. Ideas on human behaviour, including drinking, were always coloured in the direction of piety in all things.[29] As a result, many have mistakenly come to view drinking alcohol as an unhealthy habit. However, moderate alcohol use is actually associated with good health. Have I mentioned that I, for one, couldn't eat a dinner without wine for digestion? Life with wine ensures a richer, healthier experience — not to mention its delightful social, intellectual, and mood-altering benefits. As Benjamin Franklin once said, wine is "a constant proof God loves us and loves to see us happy."

(vii) Cool Fermentation

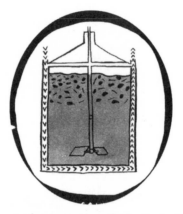

The final section in my short version of wine's significant developments over the last 2,000 years focuses on the one modern technique that has revolutionized winemaking (particularly white wines): *cool fermentation.* Until recent times, fermentation temperatures were ruled by the climatic conditions of a winery's location. If you were making wine somewhere in the Mediterranean, then as a general rule, the temperature at harvest time (September and October) would be quite warm. Higher temperatures

29   *Puritans in England:* Puritans were members of a Protestant movement within the Church of England that wanted to push the Reformation to greater extremes and stop certain practices still tolerated and inherited from the Catholic Church.

actually encourage fermentation because yeast thrives under these conditions.[30] However, hot fermentation has a major downside: The *fruitiness* of a wine is greatly diminished.

The effects of warm, quick fermentation is especially detrimental to white wines, which count on fruit flavours as their most important attribute. (Red wines have greater complexity due to fermentation on the skin and often aging in wood, so fruit flavours are, to some degree, less important.) Traditionally red wines from warmer climates have always been of acceptable quality — not so for whites, which were either bland and flat or just plain awful. Sometimes whites were even oxidized! The importance of cool ambient temperature during white wine fermentation cannot be underestimated, as it produces wine that is more floral, refreshing, and fruity. Over many centuries, winemakers came to understand that the best whites (and more-nuanced reds) mainly came from cooler regions. The challenge was how to economically replicate these lower temperatures in warmer regions of the world. Traditionally, wineries were often located in caves to allow for cooler fermentation (and storage). That was one solution.

As the making of wine spread throughout the world, especially to areas like California, South America, and Australia, the challenge of finding cooler fermentation conditions for making wine became of even greater interest. Creating underground cellars or tunneling into hillsides became common practice. Of course, modern-day use of air-conditioning in the winery was another solution. However, tunneling and air-conditioning can be expensive propositions.

The most sensible and important innovation was the use of cooler fermentation tanks, starting in the early 20th century. Cold water is run through copper coils placed in the must (the unfermented grape juice), cooling the juice. This allows the juice to be fermented at controlled lower temperatures. As the yeast converts sugars to alcohol (which also produces heat), the coils running through the fermenting tank keep the wine at a consistent temperature. The end result is wine that retains much more of its delicate

....

30 *Stuck fermentation:* There are upper limits to the ideal temperature for fermentation; as the temperature rises above 35°C/95°F, the yeast cells are killed and fermentation stops.

fruit essence. For the first time in history, great wine — especially whites — could be made almost anywhere in the world.

Today most wineries are equipped with stainless steel tanks that have cooling jackets. Each tank is made of two layers of steel, allowing for cold water to circulate around the outside of the tanks. Fermentation and subsequent storage of the wine is precisely controlled. Where weather once determined the temperature, technology now moderates. Beautiful Chardonnays that rival the great wines of France's cool Burgundy region are now made in places like California. The world of wine has changed for the better, thanks to cool fermentation.

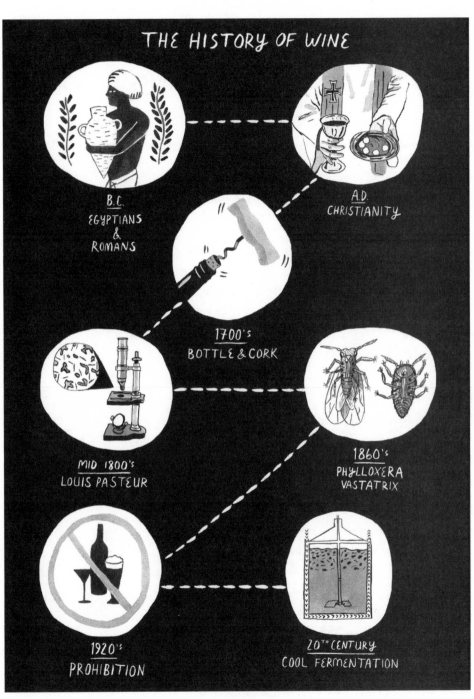

THE HISTORY OF WINE

B.C.
EGYPTIANS
&
ROMANS

A.D.
CHRISTIANITY

1700's
BOTTLE & CORK

MID 1800's
LOUIS PASTEUR

1860's
PHYLLOXERA
VASTATRIX

1920's
PROHIBITION

20TH CENTURY
COOL FERMENTATION

Fig. 10

IV. WINE TASTING

*"To enjoy wine . . . what is needed is a sense of smell, a
sense of taste, and an eye for colour. All else is experience
and pleasure."* —*Cyril Ray, English journalist*

At last we arrive at the most rewarding part of our subject: the actual tasting
of wine! I want to emphasize that you do not have to be an aficionado to
enjoy this multifaceted beverage. You only need an open bottle of wine,
curiosity, and, most importantly, a receptive nose — yes, *nose* — along with
your palate.

(i) Concentration

The one essential habit for wine appreciation is *paying attention* to the wine's
characteristics. This practice must accompany each step and every aspect of
drinking wine. Just a short few moments of concentration are all you need.
It's amazing how much paying attention will improve your experience and,
ultimately, your appreciation of wine in general.

I am not suggesting you sit around the dinner table and analyze the wine
as if you were about to write a doctoral thesis on the subject. Instead, as the
wine is poured into the glass, just take a quick look at the wine's colour. See
how the light plays with the hues at the edges of the wine. Take a nice, deep
smell. Remind you of anything? Take a sip and let the wine pass over your
tongue. Experiencing any tingly sensations on the sides of your mouth? Is
there a bitter aftertaste? Smooth? Full? Now you're getting into it — continue your dinner conversation and enjoy. This process won't make you an
expert, but you'll be surprised how much information you can collect in
those few seconds of simply paying attention to what you're doing.

Experts taste flights of wines, sometimes a hundred at a time, on a weekly basis. They usually do it midmorning, when their palates are sharpest. They spit out the wine (inebriation, though it's kind of nice sometimes, dulls our taste perception) and analyze every aspect of its character. Sounds like a good job, though I'm sure it gets monotonous. For most people, just a quick snapshot of the wine will suffice — with one big exception.

When you're in possession of a great wine, I suggest a slightly different approach. Get some friends together. Make a nice dinner, not too fancy, and *let the wine shine* — it's the star tonight! Take a break from your conversation and hosting duties. Go through the steps described above but with added concentration. Compare notes. Describe the colours and hues you see. Get everyone's opinion on the dominant smells. Take small sips. Savour the wine. Does it feel in harmony, or is there some odd, jarring sensation? Let the aftertaste linger before the next sip. Does it last or not even show up? Talk about it. Celebrate getting together and start arguing about money, sex, religion, or politics. Enjoy life. Again, move on to the meal.

That's my one exception to the rule about simply paying attention to wine. Let it dominate the conversation for a few minutes but only if it's a great wine. Otherwise, treat it like food[31] — just a part of your dinner. Wine also serves as a nice aperitif as part of any cocktail hour.

(ii) The Look of Wine

Your first introduction to a bottle of wine is its *colour(s)*. This initial experience begins when the wine is poured from the bottle into clear glass of either a decanter or a wine glass (note that most wine bottles are *coloured* glass). This moment is the perfect time to notice the complex (or not) colours of the wine. You'll see the primary colour, and then you'll see the secondary colours along the edges. The best way to observe these colours is to hold the glass at a diagonal angle. The wine in the middle is the primary colour. Now look along the area where the wine touches the glass to see the

....

31   *Wine and food:* I discuss the interplay of wine and food at the end of the book. The marriage of the two is fascinating.

hue of the wine (it's always best to do this against a white background). The colour of the outer edge of the wine can be very revealing.

As wine ages, it oxidizes and, like other fruits, turns a brownish colour. As a white wine ages, the hue turns from a fresh white-greenish colour to a golden colour. If a white wine becomes brown, then it is most likely spoiled. A red wine begins with red-purple hues when young and then turns to a brick red and then becomes pale red as it ages, finishing as a brownish pale colour. When you notice a recent vintage wine showing these brownish colours, you should be concerned about its condition.

The second attribute to observe is *clarity*. A well-made wine will appear bright and clear. Any cloudiness can usually be considered a fault — unwanted particles in suspension are probably due to poor winemaking or infection. There are two exceptions to the clarity rule. Some winemakers purposely do not filter their wine, so some sediment remains in the bottle. If you shake the wine bottle, this sediment will then become suspended in the wine, and in this case sediment is not a fault. Secondly, as great wines age, mainly red ones, they create, or *throw*, sediment due to chemical reactions over time — mostly a positive development. You can avoid this situation by properly decanting the wine. The proper decanting process begins by letting the wine stand vertically, up to 24 hours in advance, and then carefully opening the bottle and slowly pouring the wine into a decanter until you begin to see sediment or a cloudy liquid appears.[32]

(iii) Smell

It is often said that tasting wine is actually *smelling it*. Your nose, or olfactory system, can identify more attributes of a wine than do taste buds. In fact, your nose can differentiate over 5,000 different smells and is therefore your most important tool for wine "tasting."

The proper way to appreciate the way the wine smells is twofold. First, you sniff the wine quickly when it is in the glass. It's helpful to swirl the wine

....

32   *Sediment:* Do not discard the remaining wine, because it is perfectly fine. You can filter it through a coffee filter and drink!

gently to release the aromas and bouquets (the actual components carrying the smells are known as *esters*).Next take a small sip of wine, hold it in your mouth, and then draw air over the wine. This released the smells into the roof of your mouth, where the esters will enter the back channels of the nose passages. Both experiences will give you a full sense of the wine before the actual tasting. (See Fig. 11.)

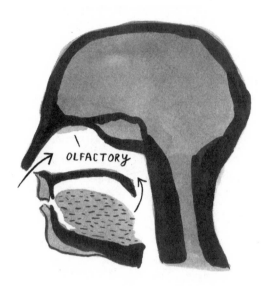

Fig. 11

AROMA AND BOUQUET

Aromas and bouquets differ from one another. The *aroma* is the smell that comes from the grape itself and the original fermentation process. Therefore, a wine made from Riesling will almost always smell the same as other Riesling wines. There will be a floral, fruity smell that is typical of all Riesling grapes. However, wines will be transformed during bottle aging, leading to the formation of new esters. These new smells are called the *bouquet*. Again using Riesling as the example, we find a very interesting development in older wines made from this grape. It will begin to smell like fusel oil! Surprisingly, this aroma is a very *positive* sign that the wine has aged well and developed added complexity. As with certain cheeses, distinctive or foul smells does not mean foul taste; in fact, in many cases, quite the contrary is true.

In summary, wine tasting starts with the visual inspection, which is then followed immediately by focusing on the fragrance of the beverage. These first two steps, to a large extent, have already foretold the character of the wine. The pleasure, however, is yet to be fully realized.

(iv) The Taste

The next step involves *tasting* the wine with taste buds as the "lens." Your tongue is covered by receptors that can distinguish only five primary sensations, or "tastes." These primary taste buds reveal only sweetness, saltiness, acidity (or sourness), bitterness, and umami (savouriness). The receptors are located all over your tongue but are not evenly distributed. Fig. 12 reveals parts of your tongue where there are stronger concentrations of certain taste buds.

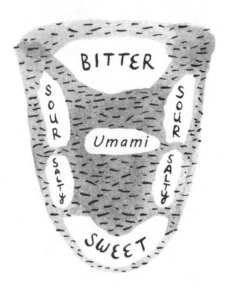

Fig. 12

Allow the wine to sit briefly on your tongue and differentiate the sensations. A wine with residual sugar will register most strongly on the front tip of your tongue. Any mineral salts will be detected mainly in the middle sides. Acidity, a feeling of crispness, will affect the sides of your tongue and mouth cavity. The bitter components, substances like tannins, will be noticed at the back of your mouth. Wines with concentrated tannins[33], like fine Bordeaux, will actually make your mouth pucker, and your teeth will feel a bit furry.

For years, most wine experts recognized only these first four tastes. However, your mouth can discern another background taste in a wine. It is called *umami,* a Japanese word that means "pleasant savory taste." It refers to the smooth, meaty sensations you get from delicious, full-tasting foods. Think of a delicious, juicy steak! You'll often find this sensation in good wine. The cause of this taste is attributed to glutamates (or amino acids) in food and beverages. This compound rounds out a wine and helps give it that total balanced flavour we all enjoy. Umami is detected by taste buds on the tongue and other parts of the mouth.

WEIGHT

The mouth also detects a characteristic called *mouthfeel,* which is the "weight" of the wine. Some wines actually taste light due to fewer flavour elements. These wines are generally white wines and young fruity reds that have no oak aging. Professional tasters describe such a wine as "delicate," "fresh," "charming," "nuanced," or "subtle" if it is properly made. On the other hand, if the wine is poorly made, they will say the wine is "thin," "lacking character," "simple," or "one-dimensional."

The more complex wines actually feel heavy and may seem to completely fill up your mouth. Typically, these wines are red, but some are aged whites made from Riesling or Chardonnay grapes. Seasoned tasters will use words like "sturdy," "weighty," "round," "spicy," "deep," "complex," or "full" to describe such a wine (if it is well-made and balanced). If describing poor examples, they might say the wine is "astringent," "dull," "flat," "cloying," "jammy," or "coarse." Many of these terms are personal and could be used for whites and reds.

....

33   *Tannins:* These substances come from the grapes, skins and stems especially, and also from the wood barrels used for aging. They help a wine age and develop in the bottle if in the right balance with the fruit.

(v) The Finish

This last aspect of the tasting experience is the one I have the fewest words for — yet I believe it is the most important. The aftertaste or "farewell" of a wine is known as the *finish*. The lingering flavour present long after you swallow the wine is often the best indication of quality. Poor wines have either little or no finish, or worse, a foul one. In contrast, great wines stay with you, so you get to enjoy them over and over again. The first time I had a bottle of Château Lafite, a top growth wine from Bordeaux, I could still "taste" its lingering flavours vividly the morning after. That is a sign of a great wine!

(vi) Wine-Tasting Strategies

Drink wine frequently if you want to truly appreciate this age-old beverage. I mean, really, what could be more delightful than enjoying a glass or two of wine on a daily basis? Few, if any, activities measure up (well, there might be one or two!). In this section, I outline five ideas for increasing your wine knowledge.

These tasting approaches work best if you create a small group of like-minded folks. Together, your group will make these tasting strategies less expensive (and more fun!). You can pool your resources so you can splurge on a great bottle once in a while. American writer and television personality Clifton Fadiman once wrote, "A bottle of wine begs to be shared; I have never met a miserly wine lover." It's worth remembering and emulating. Note that all you need is about 2 ounces of each wine per glass. Use bread and water to clear your palate between sips.

BY STYLE

One of the most common approaches to tasting and comparing different wines is to taste only one kind at each sitting. Start with tasting four different wines of the same style[34] and compare notes. For example, you might

34   *Wine Styles:* Refer to the Wine Classification Table in Section One — Wine Types and Styles.

begin with tasting complex, full-bodied reds from the same region at your first tasting. At the next tasting, try the same style — complex reds — from four different countries. You can apply the same breakdown and strategy to all the other types (other red styles, then all styles of whites, dessert wines, sparkling wines, and fortified wines). By narrowing down each category, you'll taste wines that are quite similar in many characteristics while also allowing the various differences to stand out more clearly. This approach is a good starting point for a novice taster.

BY GRAPE

This is another simple kind of tasting to organize. Go to your favourite liquor/wine store and purchase four wines all made from the same grape. Taste each one and concentrate on the differences. You'll be amazed at the varying tastes that come from the same grape. Chardonnay grapes will provide an excellent example and, for some, a real revelation. You'll find some very crisp, light wines with subdued fruit smells and flavour alongside powerful, complex-fruit, oak-tasting classic Chardonnays. Each expression is a result of the place the grapes were grown and the winemaking techniques applied, among other factors.

The first two tasting strategies — tasting by style or by grape — are easy to set up. The next two take more resources.

BY YEAR

This type of tasting will narrow down the various factors that create varying wine tastes. A good example of this strategy would be to get four different Pinot Noirs from the same vintage (year), preferably from the same area (e.g., Oregon). If you use this strategy, you'll be able to notice the small differences terroir might make . . . or the differences from grape-growing practices . . . or from the skill of the winemakers. This sort of tasting is more sophisticated and focused. One problem with this approach is that many liquor/wine stores may not have the same vintages of one type of wine on hand, and in particular, the stores often don't have many wines from the same region or older vintages.[35] Plan on visiting a few stores and give yourself a bit of lead time if this approach appeals.

BY DIFFERENT VINTAGES OF THE SAME WINE

This type of tasting really excites me. In wine-circle parlance, it's called a *vertical tasting*. This entails accumulating a collection of years of the same wine — the exact same selection from one winery. For example, get four bottles of Robert Mondavi's Reserve Cabernet Sauvignon (identical labels *but* different years) from 1970, 1975, 1980, 1985, etc. The beauty of these tastings is that you can discover how well some particular wines will age.[36]

You can facilitate this wine-tasting idea by starting your own wine cellar. When stocking your wine cellar, start "laying" down the same wine from different vintages for a few years. I suggest that you keep a log book with notes on when you bought a wine, the cost, and its potential longevity. Some people put tags on their wine racks in their cellar; a tag may simply say, "Drink before 10 years old," based on their own experience at a vertical tasting or from reading a book by a wine expert.

One of my all-time favourite wine experiences was a vertical tasting of the legendary Château Musar from, of all places, Lebanon (surprisingly, Lebanon is one of the oldest areas of continual winemaking!). This wine is made primarily of the superb Bordeaux-inspired grape, Cabernet Sauvignon, and blended with a grape commonly used in Southern France, the Cinsault. I just happened to walk into a wine merchant's premises to be told that the day before, he had led a group through twelve different vintages of the Musur spanning over 30 years. "Damn, I'm a day late," I exclaimed. My associate smiled and told me that there was a heel of each bottle left over. The 24 hours in the bottle allowed the older wines to open

....

35 *Older wines:* It may be easy to find wines from the last two to five years, but trying to buy ten- or twenty-year-old wines of the same type, vintage, and region will be a real challenge.

36 *Aging factors:* Of course, not all vintages are alike. For example, you may find the 1970 a much better wine than the 1980. This often means that the older wine (1970) came from a year that had ideal growing conditions, which resulted in superior grape production and/or quality.

up beautifully, thereby creating full, expressed flavour. The ensuing tasting was mind-blowing!

LOCAL VERSUS FOREIGN TASTING

For organized tastings, it is fun to compare local wines to other wines of the world that you've come to appreciate. This is a marvelous and tremendously enjoyable exercise. I love getting into a wine region and going from winery to winery (especially by bike). Taste and analyze and have fun. I highly recommend doing this yourself or with a group, but beware you might get hooked and end up organizing wine-themed holidays.

I have a confession to make: I am a *locavore*. Can't help it — just part of my personal evolution (soon to be a planetary revolution, I hope) . . . it's a long story. Regardless, I'm a big fan of buying local because it makes good sense for a lot of reasons — things like strengthening the local economy and reducing our carbon footprint. The beauty of learning the Three Keys introduced in this book is that, in many cases, it will allow you to purchase a wine from your own region that will match one of your international favourites. Let me illustrate my point with two examples.

Imagine a lover of white Burgundy wines who lives on the West Coast of the U.S. Traditionally these fine wines are made from the Chardonnay grape. So what are that person's options? Spend $30 or more for the French wine, or go local. No problem. There are excellent Chardonnays made in the U.S. (California, Oregon, and Washington State) and Canada (Ontario and British Columbia) to choose from. These regions produce an abundance of comparable wines that are, in many cases, half the price.

A second example relates to one of my preferences. I am very fond of reds from Bordeaux. The great examples from this French region are normally made by blending the Cabernet Sauvignon, Merlot, and Cabernet Franc grapes. I live near the Niagara region in southern Ontario, where many wineries are growing those same grapes with some success. In warmer summers, some wineries are producing Bordeaux-like reds with strong character and good quality. (Of the three grapes, the Cabernet Franc is showing the best potential; when used by itself, it often rivals the fine reds from the Loire Valley, where that grape is also used by itself.) Niagara is still a pretty young region in winemaking terms, and each year, as the grape vines mature, the product continues to improve. Now I can often get comparable Bordeaux style wines at superb value grown right here in my own neighbourhood - it's fun to taste them local against foreign.

(vii) Terms for Describing Wine

In trying to translate your sensual experiences when drinking wine, it's useful to have some basic terms in your arsenal. These words can serve as cues that stimulate and improve your perception.

APPEARANCES

Descriptors of a wine's appearance can concern a wine's clarity, its fluidity, and its hue or tint. Following is a simple list of descriptors for each of these three categories:

Clarity: *Bright, brilliant, clear, dull, cloudy* (some wines seem so clear and pristine that they seem to almost shine)

Fluidity: *watery, thin, full, viscous* (swirl a wine in your glass so it coats the sides; if the wine runs down quickly, it is probably watery or thin; if it seems to stick to the glass, it is fuller; and if it forms columns of wine, looking like tears pouring down the sides, then it has high viscosity — these "tears" are also known as the *legs* of the wine and usually indicate high alcohol)

Hue or Tint: The colour of each wine type can be broken down as follows (the colours progress in order from that of a young wine to an older one):

> **Red:** *Purple, red, dark red, cherry, garnet, light brownish, brown*
> **White:** *Greenish, yellow-white, lemony, straw, gold, amber*
> **Rosé:** *Reddish, pink, salmon-grey, orange, light brown*

SMELLS

The University of California has developed a guide known as The Aroma Wheel (see Appendix #4). It is particularly useful in providing cues to help you identify particular and precise smells. For example, if you think a wine smells "spicy," simply go to the wheel's inner circle and follow *spicy* to the edge of the graph; you'll find *cinnamon, cloves, black pepper, anise,* and *mint.* As your nose gets trained to identify smells, you'll get better and better at coming up with the exact descriptor.

TASTES

The third part of your tasting vocabulary relates to the five sensations your tongue can identify — sweet, salty, sour, bitter, and umami (savoury) — along with overall mouth weight. The main goal for your palate is to determine whether the wine is balanced (i.e., it has no overly dominant flavour).

Sweet: *Dry* (meaning no sugar), *off-dry, fruity, sweet, very sweet*

Salt: *Neutral, mineral, salty* (overall, wines rarely taste very salty)

Sour (acid): *Flabby* (little or no acidity), *crisp, tart, sour*

Bitter: *Flat, soft, astringent, bitter* (this character usually is a result of the amount of tannins in the wine; the more tannins, the greater the bitterness)

Umami: *Simple, tasty, flavourful, savoury*

Weight or Body: *Thin, light, medium, big, full, heavy*

Note on the finish or aftertaste: The all-important after effects of drinking wine — its finish — can be described basically as *short,* meaning there was none; *medium,* when the wine's taste disappeared fairly quickly; *long,* denoting a lingering aftertaste; and *powerful,* describing those great wines that seem to be just as impressive in their finish as they were in their original smells and tastes. You get to enjoy those wines with a powerful finish minutes to hours after drinking the wine!

CONCLUSION TO
SECTION ONE: THE BASICS

I hope that this first section provides a good foundation upon which your wine knowledge can grow and flourish. Our discussion of the basic winemaking techniques can help you understand some of the challenges associated with viticulture and vinification. Not only is the cultivation of grapes the beginning stage in winemaking, but it's also the most important in many ways. Just as poor ingredients limit the quality of a great chef's final dish, regardless of techniques and fancy sauces (which should enhance the food, not disguise it!), wines are limited by the quality of the grapes.

We have discussed the fact that although turning grape juice into an alcoholic beverage is a natural phenomenon, winemaking (*vinification*) is fraught with many unexpected problems. For example, poor sanitation leads to infections by bacteria or other microorganisms that ultimately spoil the wine. Understanding what is truly involved in making a wine will help you more fully appreciate the actual tasting experience.

Knowing the different wine types and styles will allow you to distinguish wines, one from another, and develop your favourite categories. The wise old Greek philosopher Aristotle famously broke knowledge into its elements, or simple parts, to increase understanding. His use of "categories" became the model for much of modern analytical thinking. Knowing the types and styles of wine will allow you to compare apples to apples. You can now take similar wines from different regions or countries that are defined as one style and compare them. This will help you learn about the variations caused by grape type, climate, techniques, and so on and thereby increase your understanding and enjoyment.

Of course, nothing can be quite as illuminating as studying history. It's fascinating to find out how a certain wine arrived on the scene. I love the story of how the famous monk Dom Pérignon "captured" the bubbles in champagne. Myth or truth . . . who really cares? The stories themselves enrich our experience. And for the professionals making the wine, studying history is imperative. The old saying "those who ignore history are doomed to repeat it" is especially good advice in this business. Learning and adapting from stories of the past, from experience, and from one another should always be the winemaker's credo.

Finally, the last part of this section covered the rules of engagement for drinking this magical beverage. This understanding will give you the confidence to sit down with either novice or expert to taste and discuss a wine's attributes. For simple enjoyment, wine should be sipped with only passing attention — even quaffed in some cases (especially with young wines like Beaujolais Nouveau). To truly appreciate a fine wine, however, we need to take a more measured approach. You now have the knowledge and the terminology to express yourself effectively.

Time to move on to the atomic particles of our beverages: the grapes.

SECTION TWO: THE GRAPES

I. INTRODUCTION TO GRAPES

"Wine . . . the blood of grapes." —Genesis 49:11

The quality of the grapes is the primary element in the eventual outcome of a wine — good or bad. The condition of the grapes and their degree of ripeness is critical to the best possible final product. No amount of vintner tricks or alchemy can overcome the lack of quality from the vineyard. Aside from good grape condition, the other critical factor is the type of grape. Which variety is being used?

There is general agreement among experts that only a handful of grapes are great — or to use the fine wine writer Hugh Johnson's term, *noble.* The point is that only these few varieties, alone or in combination with other noble varieties, will result in superior wine. These are the wines that have pleasant and/or more distinctive aromas, complex flavours, and, quite often, greater longevity.

(i) Eight Classic Grapes

As you know by now, memorizing facts and figures is not the purpose of this book. (A better option is to use reference texts and other sources to look up particular information as you need it.) However, knowing the names of the eight classic grapes is worthwhile. These few varieties account for approximately 75% of the world's quality wines.

The following grapes certainly represent most of the classic grapes by majority opinion, although their greatness is always open to some dispute:

> ***Classic White Wine Grapes:*** Chardonnay, Riesling, Sauvignon Blanc, and Gewürztraminer

Classic Red Wine Grapes: Cabernet Sauvignon, Pinot Noir, Merlot, and Syrah (Shiraz)

(ii) What Makes a Classic Grape?

All of these classic grapes belong to the vine species known as *Vitis vinifera.* This family of vines includes all the great wine grapes without exception. There are other species of the *Vitis* genus, but they do not produce fine wine. However, some varieties in North America were used for making wines in the early days of colonization. Their main contribution was the role they played in resolving the devastation caused by the insect phylloxera (see the history of wine in Section One). That louse ruined the best vineyards in Europe but found its match in an American native rootstock.

Besides being members of the *Vitis vinifera* family, the classic grapes have various characteristics that, in appropriate quantities and with proper care, give them the potential to make superior wines. These components tend to give these grapes definitive character — and separate them from the crowd. Their characteristics are widely recognized as producing better wines than those derived from other grapes.

It would take a scientist to identify all the individual components that help to elevate the classic grapes above the rest. In Bordeaux and California, you'll find the two top universities that are at the forefront of wine research. They are devoted to analyzing and understanding all that goes into the making and appreciation of wine. You can read countless treatises on grape chemistry and attributes.

For our purposes, you simply need to know the four key elements to white and red wines (rosés too!): sweetness (fruitiness and sugar), acidity, alcohol, and tannins. These four factors determine the aromas and bouquet, initial taste, mouthfeel, and aftertaste or finish of a wine. The secret to the great wines is that these four elements are in *balance.* The classic grapes most frequently deliver this desired outcome.

(iii) Other Important Grapes

In addition to the eight varieties that account for 75% of the world's quality wines are nine that some people include in the "classic" or "near-classic" grouping:

> **White Wine Grapes:** Chenin Blanc, Muscat, Pinot Gris (Grigio), Semillon, and Viognier

> **Red Wine Grapes:** Nebbiola, Sangiovese, Gamay, and Tempranillo

We'll discuss these pretenders to the crown later, along with some other common grape varieties and some of the ancient grape varieties.

It might be more an accident of history than good planning that the "classic" grapes became the chosen few noble varieties. For example, the ancestor of the Riesling in Germany was likely a native grape from that region. When the Romans initially colonized that area, they probably made wine from that wild variety. Over time, winemakers would have practiced selective breeding, taking the best plant cuttings for new planting. They would have also cross-bred the grape with more familiar grapes that they imported into the region. Over the centuries, the modern Riesling appeared — one of the classic, great winemaking grapes on the planet. The same sort of accidental or coincidental history can likely be attributed to the other classic varieties.

This raises an interesting question: Have other grape varieties been ignored? Are there grapes growing somewhere that if planted elsewhere might be classic in character? It doesn't take a great leap of logic to imagine that other grapes from the ancient world could have proved to be noble. Perhaps indigenous varieties still lurking near the Mediterranean, in Syria or Persia (Iran), for example, would have flourished in more northerly climates. Maybe this is a tale yet to be told!

(iv) Historical Benchmarks

In ancient Greece, the great philosopher Plato proposed the concept of an *archetype*. This term referred to an ideal model upon which all things are based. For example, a table would be patterned upon some original principle or prototype of all tables. Whether these archetypes actually exist in some way is open to a lot of debate. I think that at times, Plato argued that they actually do exist in some heavenly realm. Others would say that these archetypes could be part of our psychological foundation or the unconscious (perhaps something to be discovered in the "collective unconscious mind" as put forward by Carl Jung?) . . . all interesting stuff, but for our purposes, only the basic concept is important.

An archetype is what we'd also call a *benchmark,* something against which all subsequent examples of a certain item or thing can be measured. In the following pages, we'll discuss the historical benchmarks of the classic grapes. Why? First, virtually all winemakers will strive to *emulate* the great wines. They want to chase the classic expression of that grape found in the area wherein it initially reached its pinnacle. In the same way, a violin maker tries to match the beauty and sound of a Stradivarius. The vintner will grow his grapes in a time-tested fashion, pick them at a prescribed ripeness, vinify using proven practices, and age the wine in a standard manner and for a set amount of time. All this is done to replicate the benchmark wine. For example, some Australian winemakers will try to make their Shiraz (Syrah) match the wines from the Côtes du Rhône in France.[37]

The second reason for using benchmarks is that it gives us a common language. The archetype concept allows us to compare the old with the new. The winemaker in California can say her Cabernet Sauvignon tastes just like a Medoc from Bordeaux, France. Sometimes, and this is when it gets interesting, that same winemaker or an industry expert will say a wine is superior to the former benchmark. The gold standard has been moved! It gets even more exciting when people say the wine is *different* and better. One example in recent times is Sauvignon Blanc.

Traditionally grown in France in the Loire Valley and Bordeaux, the Sauvignon Blanc grape creates wine that is complex, exhibits grass-like

....

37 *Australia:* Generally, Australians don't try to match the French. It is not in the Australian character! They have traditionally created "fruit bombs" out of Syrah (Shiraz), which taste more like juice than wine. Recently, Australian winemakers started making subtler and more nuanced styles, similar to the French.

aromas, is quite dry on the palate, and is not generally known for its fruiti-ness. Now Sauvignon Blanc is also grown in New Zealand, where it commonly displays exotic fruit aromas and flavours. In better examples, it also has some of the grassy complexity you'd find in France. The New Zealand version can be an all-around superb tasting wine that in some ways exceeds the French wines made from the same grape. Other examples will surely arise as winemaking techniques improve around the world and the New World wine regions strive to meet, and ultimately exceed, the Old World benchmarks.

Back to the original point: The whole reason behind the historical benchmark is that it gives us a starting point for comparisons, allowing us to better communicate about wine attributes when discussing the pros and cons of wines from different regions.

II. WHITE GRAPES

"Montrachet should be drunk kneeling, with one's head bared." —Alexandre Dumas, French novelist

(i) Chardonnay

The Chardonnay is undoubtedly a classic grape. Many would argue it has always been the best white wine grape in the world. However, Chardonnay is also the grape that has had the most curious recent history. For centuries, the Chardonnay variety produced quality wines that writers extolled time and again. The greatest Burgundy whites are made from this grape, yielding wines with charming aromas and bouquets, complexity, and long-lasting flavours. Yet it is also a much-maligned grape, due for the most part to its constant presence everywhere. It was ubiquitous, almost annoying! And in the mid-1990s, it gave rise to the now infamous ABC movement — Anything But Chardonnay. You could not escape a glass of Chardonnay at most social gatherings.

One of Chardonnay's great attributes has also proved to be somewhat of a curse. The "problem" is its versatility and traveling genes. Because the grape is strongly affected by soil types, Chardonnay will make different tasting wines in various regions. Furthermore, a skilled winemaker can create any number of final expressions of the grapes, using distinct techniques like aging in oak barrels. Wine made from Chardonnay can range from neutral, lean, and acidic to aromatic, full-bodied, and fruity. In between, you'll find wine with a nuanced bouquet, buttery vanilla flavours, and long-lasting pleasant aftertastes. The list of styles is staggering. Unfortunately, many Chardonnays are also inferior and unflattering. Over-oaked and fruit forward, fat, and flabby are also common styles of Chardonnay, as are wines that have no character at all. As a result, people's appreciation of the grape was sidetracked for a time.

The second "problem" characteristic of the Chardonnay is its ability to grow under almost any climatic conditions. It travels well, meaning it can reach ripeness in many different regions and still produce quality wines. A Chardonnay-based wine is available virtually everywhere wine is produced. And as noted earlier, everyone served it at social gatherings. The prevalence of Chardonnay was compounded by the fact that in the last part of the 20th century, the preference for white wines outnumbered red wine by 2 to 1 in North America, and Chardonnay was the most popular wine among those whites. People simply got tired of tasting the wine. A pity!

Historical Benchmark: Chardonnay's history dates back to at least the 14th century in France. Its name may have come from a small village in the Mâconnais region of Burgundy. Near that area, the grape reached its apex, thanks in large part to the various orders of monks who lived in the region. For hundreds of years, the wine had been used for religious purposes and, you can be sure, for pure enjoyment. There is little to compare to the pleasure associated with drinking a Puligny-Montrachet or Meursault from Burgundy.

The two primary (Chardonnay varietal) benchmarks are easily established by breaking Burgundy into north and south regions. In the north, Chablis country, the Chardonnay wines are made with a delicate aroma, in a crisp, lean mineral style, and seldom with any oak contact/influence. The southern Burgundy style, which is naturally fruitier than Chablis and uses oak barrels, has a more pronounced aroma and greater bouquet. The smells and tastes range from apple, pear, and melon to nuttiness, butter, vanilla, and herbal — complex in almost every way. Either way, the wine is exemplary and is considered to be the one white wine that is closest in character to reds. (You'd be surprised how many people can be fooled in a blind tasting of Chardonnay — blindfold required so the colour isn't noticed.) The whole world looks to Burgundy for comparison and the measure of their Chardonnay wine.

Note on Champagne: The great sparkling wines from Champagne are usually made from two to three grapes. Two of them are red, Pinot Noir and Pinot

Meunier. The other grape is the Chardonnay — it is considered the best white grape for making bubbly wines. When grown in northern regions, the grape maintains a pleasant acidity, which is always required for making a good, balanced sparkling wine.

CHARDONNAY FACTS

Colour:	White (greenish skin)
Grown:	Worldwide (thrives in cool and moderate climates)
At Home:	Burgundy, Champagne, California, Australia, Chile, Canada (almost everywhere!)
Soil:	Limestone, chalk
Character:	Crisp, acidic, fruity, green apple, vanilla, round, nutty and complex

(ii) Riesling

Many people, including the English wine expert Jancis Robinson, think the Riesling grape rivals the Chardonnay as the greatest white grape. Yet Riesling also suffers from neglect, similar to the often-maligned Chardonnay, though for a different reason. Riesling is a "forgotten" and misunderstood grape in North America; it's generally dismissed as "too sweet" and so is served infrequently. What a mistake!

The Riesling variety produces some of the longest-lived white wines in the world. These wines can age for 20 to 30 years. Its very best examples manage to deliver floral bouquets and terrific natural fruit flavour with citrus overtones, and they finish with a crisp, clean, and refreshing lingering aftertaste. This delicate balance of smells, fruit, and acidity is a marvelous combination — a work of art. Drinking a great Riesling wine is like experiencing the first warm day in spring overlaying a nice remnant of a winter's cool breeze: It's refreshing and heartwarming in the same glass.

The Riesling grape vine is one of the hardiest of root stocks. This characteristic allows it to grow in northerly climates and delivers its finest wines in these areas, where ripeness is often barely achievable. The old truism about the best wines coming from climates and soils where the grapes must struggle to reach ripeness is especially true of the Riesling. The cool climate results in higher acidity, which is the backbone of this wine and the key to its crisp flavour and balance.

You would think that a vine that can grow under such difficult conditions, in cool climates and often stony ground, would invoke some sort of mythology that would create demand. Think of other foods that grow only in limited locations or circumstances (certain teas, mushrooms, fruits, etc.) and the associated value placed upon them, sometimes just for their tenacity and scarcity. In this age of persuasion, you'd expect the marketers of Riesling to tell this narrative. The wine would do the rest!

Historical Benchmark: The Riesling grape has been grown in Germany for over a thousand years. There is some speculation that it is a cross between a wild grape native to the region and a Mediterranean grape brought to Germany by the Romans. The first mention of the variety dates from 1435 in property records of a principality on the Rhine River. By 1477, it was also growing in Alsace, France. Regardless of its parentage, the Riesling has a long history, and its stylistic variations are well-established.

In terms of style, Riesling has a range defined more by sweetness than flavour (that's not to say it isn't packed with flavour). The two archetypical versions are German and French (though the French is from Alsace, which is an area that once was Germanic). In Germany, the wines are almost always found with residual sugar. This doesn't mean the wine is actually "sweet." Grapes grown in northern climates (and inherently in some grapes, like the Riesling) are higher in acid. Not only does this give the wine structure, or backbone, but it also renders the sugar neutral in a way. Riesling arrives in your mouth full of sweet fruit floral aromas and tastes but exits dry with a complex, lingering finish that still has a delightful fruity character.

In France, specifically Alsace, the Riesling is mostly given a drier treatment and overall leaner expression. This wine deserves a meal accompaniment. Like some German Rieslings, Alsatian variations will age well for a decade or more. The combination of fruit extract, some sugars, and high acidity makes the Alsatian a rare wine that dances between fruit and complexity

— imagine a steely dry Chablis (Chardonnay) with an eye drop of intense floral and fruit extract! That's an Alsatian Riesling.

A final word on Riesling's attractiveness is in order. In this world of high-alcohol "muscle" wines, the Riesling offers a needed respite. Here's a wine that will often come in at less than 10% alcohol yet be still packed full of flavour — fully one-third less alcohol than the many Robert Parker–influenced alcohol monsters. Riesling is wine that is interesting, complex yet delicate, and low on the booze scale. Alleluia!

Note on Very Sweet Rieslings: Riesling is a grape that also lends itself to what we'll call "dessert" versions. The three main types are:

- *Late harvest:* This Riesling is a result of leaving the grapes on the vine well into the fall season. The sugar continues to rise and reaches levels that leave residual amounts even after fermentation.

- *Botrytized wines:* These wines are a result of the grape being attacked by a fungus, *Botrytis cinerea,* the so-called "noble rot." This fungus dehydrates the grape, thereby concentrating the sugar.[38]

- *Ice wine:* Ice wine is the ultimate high-sugar style of this grape. The grapes are left on the vine until they are frozen solid and then are pressed while still frozen. The result is that the sugar of the grape juice is more concentrated, while most of the water remains frozen. The ice wine is a rich elixir, a dessert unto itself!

RIESLING FACTS

Colour:	White (gold, yellow)
Grown:	Worldwide (thrives in cooler climates)
At Home:	Rhine, Mosel, Alsace, New York and Oregon State, Australia, New Zealand, Canada

....

38 *Sugar Levels:* In Germany, the best white wines are classified as Quality Wines (Qualitätswein) if they reach a certain level of sugar naturally. If they go past a certain base level, they are called, in ascending order of grape sugar content, Kabinett, Spätlese, Auslese, Beerenauslese, and Trockenbeerenauslese.

Soil:	Rocky, slate, sandy
Character:	Floral, honey, fruity, peach, acidic, sweet, mineral (petrol notes in older wines)

(iii) Sauvignon Blanc

In the introduction of this section, I mentioned how different New Zealand Sauvignon Blancs are from the typical French wines made from this grape (where it found its original expression). Both versions share a grassy taste and gooseberry aromas.[39] However, the French style tends to show certain almost rotten bouquets — think wet wool (and cat pee!) — while the New Zealand varieties are bursting with exotic fruits and incredible freshness. It's as if the New Zealand winemakers are using a different grape altogether. Maybe it's the grape's character?

The root word for the grape's name is *savage* in English (*sauvage* in French). The name derives from the vine's tendency to grow vigorously. This grape needs to be managed carefully in the vineyard to create a good wine. If it isn't handled with care and the growth isn't managed, or if the growing conditions are not good, then the wine tends to be thin, lacking in complexity and fruitiness. Perhaps the difficult character of the grape, or its wild streak, explains its tendency show different faces in the final wine.[40]

Sauvignon Blanc has experienced a period of tremendous growth in popularity in recent years, displacing Chardonnay as the go-to wine at many social gatherings. This development is in large part due to the fresher and fruitier expressions from the New World, particularly New Zealand. This version of the wine makes for an excellent aperitif all by itself and is also an excellent food wine due to a high level of acidity, which cuts through richer sauces and strong tastes in some cheeses.

....

39   *French version:* The French also make sweeter expressions of the Sauvignon Blanc grape, especially when it's blended with the Sémillon grape to make Sauternes.

40   *Parentage:* Interestingly, this grape was crossed with the Cabernet Franc long ago to produce one of the greatest red grapes, the Cabernet Sauvignon.

Historical Benchmark: I'll consider the traditional style to be the benchmark of this grape's wine. The great wines from the Loire Valley near the villages of Pouilly-Fumé and Sancerre are classic wines. It's not often that wine has such distinctive, sometimes peculiar smells,[41] herbal tastes layered with restrained fruit, and a lingering, very pleasant aftertaste. That is what the great examples of Sauvignon Blanc have to show. They offer a wild ride in the wonderful world of wine.

SAUVIGNON BLANC FACTS

Colour:	Greenish white/yellow
Grown:	Many countries (thrives in cooler climates)
At Home:	France, California, Washington, New Zealand, Australia, Chile, South Africa
Soil:	Limestone, sandy, chalky
Character:	Herbaceous, fresh-cut grass, gooseberry, smoky, flinty, exotic fruits, wet wool (sometimes oaky when rare barrel-aging is employed)

(iv) Gewürztraminer

Here's a grape that should be featured in a soap opera. You either hate or love it; few people fall in between. This grape yields wine with truly distinctive character. It has highly aromatic smells with flavour to back it all up. The final grape of our classic whites, Gewürztraminer, is a controversial member of that exclusive group.

Gewürztraminer originates from mid-European regions. It is probably a descendent of the Traminer grape, which may in turn have originated in either Austria or northern Italy. At some point, the Traminer was crossed with another variety and became the Gewürz-Traminer, which literally means "spicy" or "perfumed" Traminer. Its aroma is characterized as lychee

---

41 *Smelly cheeses:* Sauvignon Blanc wines from Sancerre, especially older ones, often have a cat pee smell. Sound disgusting? Think great smelly cheese! Delicious.

fruit or grapefruit with rose petal overtones — an odd combination. The grape is naturally high in sugar, giving it a pleasant fruit flavour, which includes the tastes of citrus and a variety of spices thrown in to the mix. The wine is usually lower in acid, which makes it easy to drink on its own (i.e., without food). It often overpowers foods,[42] but one food pairing made in heaven is smoked salmon and, as it's often called, a glass of Gewürz (pronounced guh-*vourtz*).

Historical Benchmark: The Gewürztraminer grape has found a home in many cooler-climate countries, but its real home is Germany, along with Alsace, a part of France that often belonged to Germany over the years. These two countries have established the grape's benchmark. In German hands, the wine is often sweeter and more straightforward, more generic, with less character. You often find this style in many other countries that try to make wine from this grape. However, I believe Gewürztraminer has found its true home and reaches its pinnacle in Alsace — that is, if you like it! Gewürztraminer is the second most planted grape, next to Riesling, in Alsace. The wine it produces is dry yet fruity, full of smells and flavours. As it develops in the bottle, the wine's bouquet becomes a little less pronounced on the flowery spectrum, but the lychee-citrus smell stays pronounced. Some Gewürz will age very well, especially late-harvest styles.[43]

This grape variety doesn't travel too well (i.e., produce similar quality in other regions). It is fussy about soil types and takes special care to grow properly. Another reason we see little of it in other countries is its name, I think. It is too hard to pronounce, so many people won't buy or order it in a restaurant so as not to be embarrassed. Poor old Gewürztraminer — some people won't even speak its name.

....

42 *Food matches:* This wine often goes with food that no other wine will have a place. For example, it does well with Chinese food and certain spicy foods like curry which normally only match well with beer.

43 *Alsace history:* France and Germany were often at war in the last two centuries. Alsace changed hands four times in 75-year period! The native language of Alsace is similar to that of Germany. Today, most Alsations speak French, but over 40% still speak the ancient Germanic dialect.

GEWÜRZTRAMINER FACTS

Colour: Pink to light red

Grown: Few countries in quantity (thrives in cooler climates)

At Home: Alsace, Germany, Austria, New Zealand, United States Northwest

Soil: Limestone, clay

Character: Strongly aromatic, flowery (roses), grapefruit, lychee, oily, spicy, fruity

(v) Other Important White Grapes

Among white grapes, Chardonnay, Riesling, Sauvignon Blanc, and Gewürztraminer are widely considered the very best producers of fine wine. Yet, as we discussed in the introduction of this section, one never knows whether this esteem is just an accident of history. It's not hard to imagine that the big four originally became prominent for various, often arbitrary reasons that were not necessarily related to quality. After these grapes were accepted into the mainstream, they were nurtured to great heights by winemakers who focused almost exclusively on this small group while ignoring the rest. Who's to say that some long-forgotten grape could not have become a classic variety? There are, without doubt, many other fine white grapes — pretenders to the crown. Following are my five picks.

CHENIN BLANC (PINEAU DE LA LOIRE)

Chenin Blanc grape has a reputation for making rather bland, sometimes overly acidic wine. Don't be fooled. In the hands of a good winemaker, it can deliver a marvelous fruity wine with beautifully balanced acidity. Wine expert Jancis Robinson includes it among her favourite whites. The grape excels in the middle part of the Loire Valley region in France — and not only as a dry white wine. It makes an excellent sweet, fruity, late-harvest wine that finishes dry at the back of your throat. The grape is also susceptible to the noble rot, *Botrytis cinerea,* resulting in raisiny fruit that yields intensely sweet wines tasting of honey yet still balanced and not too sappy. These dessert wines are known to age well and for many years.

The grape is widely planted in South Africa and California. These wines can be quite good but are generally uninteresting with a few exceptions. For the great Chenin Blanc wines, buy them from the Loire. I have been fortunate to spend some time in the Loire Valley, mainly around the town of Vouvray, where I became quite fond of this grape. Not only did I find the sweeter wine interesting, but I also developed a fondness for the sparkling incarnations known as *Crémant de Loire*. These wines are a great value and a good replacement for Champagne (for those of us on a budget!).

MUSCAT (MUSCATEL OR MUSCATO)

The Muscat grape is widely planted in Europe and is known for its distinctive aroma: a very strong "grapey" smell. It is a very versatile grape used for dry, sweet, fortified, and sparkling wines as well as for eating.

This grape is used extensively in Chile for dry table wines and less so in Italy, where it's called the *Muscato*. In many countries, the grape is allowed to stay on the vines to maximize its sugar content. It makes a nice dessert wine that pairs well with strong cheeses. Around the Mediterranean, it is often fortified with alcohol to make excellent aperitif or digestive wines (it is also used in the blend of grapes for some Sherries). Australia is also well-known for its fortified Muscat. The Muscat also makes some pleasant sparklers in Asti, Italy.

This grape is thought to be one of the oldest known grapes of the *Vitis vinifera* family. It is actually the ancestor of many of the well-known classic varieties, both red and white. One peculiar characteristic is that the white varieties are very high in antioxidants, making them comparable to red grapes. I personally love some of the sweet Muscat wines from Southern France and the state of Victoria in Australia. The grapes are great to just eat on their own; they go from white to black in colour, with each shade being delicious. Overall, Muscat is an interesting and important grape.

PINOT GRIS (PINO GRIGIO)

Here's a grape with a double identity. The Pinot Gris (or is it *Grigio?*) belongs in a James Bond film. At one moment it's there, and then it's gone — the flavour and complexity, that is.

This grape is called *Pinot Gris* in France, where it makes grey-pink-blue coloured wines that are dry, complex, spicy, and floral. This is a great wine to have with food. Pinot Gris wines also have good cellaring potential, especially the top ones from Alsace. This grape also produces some full-bodied balanced wines in Germany.

When the label says *Pinot Grigio*, the Italian spelling for the grape, something strange happens: With very few exceptions, this incarnation is light-bodied, almost colourless, and one-dimensional on the palate. Often mass-produced and available in 52-ounce bottles, this wine is good for quaffing and nothing else. In the New World (the Americas in particular), the Pinot Grigio was popularized by the television series *Sex and the City*. The wine became a sensation, soon displacing Chardonnay and Sauvignon Blanc at fashionable cocktail parties. Call me old-fashioned, but I'll take the version with taste. Life's too short to drink trendy but boring wines!

SÉMILLON

The Sémillon grape rarely appears alone. It's commonly blended with other grapes, particularly Sauvignon Blanc. At one time, Sémillon was one of the most widely planted grapes in the world (South Africa and Chile planted it extensively). Its popularity was largely due to its abundant yields at harvest time. It was easy to grow and produced lots of fruit. At its best, Sémillon rivals some of the classic varieties.

Now this variety is largely confined to France's Bordeaux region and the Hunter Valley in Australia; production elsewhere is not significant. The French tend to blend it with the Sauvignon Blanc and sell it as an ordinary table wine, where it contributes a nice mineral and citrus flavour and a clean, grassy, floral nose. In Australia, Sémillon was known as the *Hunter River Riesling* and was used by itself to make a sweetish style of wine or blended with the Sauvignon Blanc and/or the Chardonnay (Australian winemakers are known for their break-all-the-rules tendency and often make unorthodox blends . . . which sometimes work!). However, the greatest fame accorded to the Sémillon is as a result of its susceptibility to our old friend, the noble rot.

Sauternes: In the Bordeaux region of France, the Sémillon grape takes on a starring role. It is allowed to stay on the vine into late fall, when it begins to rot because of the fungus *Botrytis cinerea*. As the fungus attacks the grapes, they lose moisture and shrivel up. Sugar gets concentrated in the grape, which in turn yields a complex, syrupy liquid. The resulting juice is blended with Sauvignon Blanc and fermented into a Sauternes.[44] The result is an amazing elixir, one that will age for a long time while presenting multifaceted smells and flavours. The most famous example of Sauternes comes from Château d'Yquem in Bordeaux. Be prepared to get a few friends to help with the $1,000 price tag! It's a once-in-a-lifetime experience — try it with foie gras. My one time was, and still is, memorable.

VIOGNIER

The Viognier grape was not very common thirty years ago. It is susceptible to disease and is often a low producer, which may well explain its rarity. Once, the only significant plantings of the grape were in the Rhône Valley region of France. But Viognier has become a fashionable grape in more recent times, and planting has increased in France, South and North America, and Australia. Why? Because of the grape's wonderful smells, deep yellow colour, and complex, fruity tastes.

Viognier produces a unique aroma that brings to mind apricots, peaches, and floral notes. It is an alluring combination. On top of the lovely, almost tropical smells is a wine that is high in alcohol and full of flavour. It has become the darling of many wine lovers who like to surprise friends with something different.

The grape is also one of those rare whites that are blended into a red wine. In the Rhône region, Viognier is added to the wines of Côte-Rôtie, where the Syrah is the predominant red grape. I've also had Australian wines where the winemakers followed the same practice using their Shiraz grape (*Syrah* in France). In most cases, 10% of the Aussie blend is the Viognier. The grape helps to stabilize the colour of the Syrah and adds alcohol to the final wine (the Viognier ripens earlier than the Syrah and will reach higher

----

44   *U.S. Sauternes:* Do not confuse French Sauternes with the generic "sauternes," which can have a variety of styles — none of which approach a true Sauternes.

sugar levels, thereby producing higher-alcohol wines). Viognier is also blended with other white grapes, such as the Chardonnay, in some blended white wines.

III. RED GRAPES

*"I had to cook a dinner glorious enough to complement
the Lafite. It took four days . . ." —Gael Greene,
Insatiable Critic, "Ma Vie Avec Le Grape Nut"*

(i) Cabernet Sauvignon

Since ancient times, the word *aristocracy* referred to the best of any class of citizen groups. Plato and Aristotle argued that people's particular skills should determine their occupation. They believed that a person best suited to govern should be among the ruling class, an aristocrat. That was the original meaning of the word.

Over time, *aristocracy* came to have more negative connotations. It now stands for a sense of privilege. People inherit the right to their position in society with no regard to merit or ability. In many societies, the aristocracy soon became a hated class. This negative attitude towards the aristocracy played a major role in some of the great winemaking areas in France (we'll discuss this in Section Three of the book).

What does all this have to do with Cabernet Sauvignon grapes? Well, of all grapes in the world, this one truly belongs to the aristocracy . . . in the very best sense of the word. It is truly a *noble* variety.

Cabernet Sauvignon hales from the Bordeaux region of France. It is used mainly in blends (with Merlot in particular) but is usually the dominant grape. It contributes a clear set of sensations, aromas, and flavours that are unique to the variety, often summed up in one word: *structure.* In other words, the Cabernet Sauvignon grape brings beautiful architecture to a wine. Imagine a wine as a great building; the Cabernet Sauvignon represents the infrastructure (framing, plaster, plumbing, electrical, etc.), while the other blending grapes, like the Merlot, bring the decorative aspects. Yet Cabernet Sauvignon is also a grape that can be the whole building. California proved that the Cab Sauv, as it is often called, makes a great wine all by itself. It is truly a flexible, and adaptable, superior grape.

This variety originally reached its pinnacle in Bordeaux, on the so-called Left Bank of the region, in gravelly soils. The vines thrive in the moderate maritime climate. Cabernet Sauvignon produces its most desirable characteristics when the seasons are warm and sunny. The immediate flavours are caused by the tannins extracted from the skins during fermentation. The typical Cab Sauv smells of black currant, and other dark berries show through. The taste yields pronounced fruit flavours and a pleasant spiciness. In a well-made example of Cabernet Sauvignon, all of these characteristics lead to a beautifully balanced wine that finishes well and serves as a great accompaniment to many full-tasting foods.

Like the Chardonnay grape, Cabernet Sauvignon is a good traveler. It loves warm, sunny seasons, so unsurprisingly, this grape does very well not only in California but also Australia, Chile, and Italy. It achieves excellent results in the microclimates of these regions and in other countries where the warm days are accompanied by cool nights. The northern regions of California, especially the Napa Valley, are perfectly suited for Cabernet Sauvignon. Some people argue that the wines of this grape made from grapes grown in California have actually surpassed those of France!

The whole France/California competition began as a result of a very famous wine tasting. In 1976, a wine competition featured top wines from California, Burgundy, and Bordeaux. The judges were mainly French, but the results represented a turning point in the entire world of wine. One winner, a California Chardonnay, was rated the top among the white wines, beating out the Chardonnays from Burgundy. And even more surprisingly, a California Cabernet Sauvignon came out on top, trouncing the red Bordeaux wines (from blended grapes but made mainly with Cabernet Sauvignon). The competition, famously known as The Judgment of Paris (where the tasting was held), shocked a lot of wine traditionalists who believed all French wine was superior to any New World wines (those same people think French wines are superior to any wines from Europe, too!).

The good news for wine lovers is that good Cabernet Sauvignon can be had from many parts of the world. The grape tends to be quite similar in character no matter where it grows (especially in similar climates and soils). This trait — uncommon in other grapes — is part of Cabernet Sauvignon's nobility: It not only travels well but also is consistent in character. Chardonnay, on the other hand, can show many faces and is more open to soil, grape grower, and winemaker influences. Cab Sauv is definitely the best in class. Iconic.

Historical Benchmark: Unquestionably, the Cabernet Sauvignon from Bordeaux, especially the Médoc area, sets the gold standard in red wine. This grape is said to have been planted in the region since Roman times. Today, we must also include California's Napa region's expressions as benchmark examples of what a winemaker can do with this grape. Take the best examples of Cabernet Sauvignon wines from either the Médoc or Napa and compare them with any others around the world; few would dispute Médoc or Napa's place at the top.

CABERNET SAUVIGNON FACTS

Colour:	Black
Grown:	Worldwide (thrives in moderately warm climates)
At Home:	France, California, Australia, Italy, Chile
Soil:	Gravel
Character:	Tannic, black currant and other dark berries, complex, vegetal, bell peppers, long finish

(ii) Pinot Noir

Pinot Noir goes by many descriptors, not all of which are complimentary. Here are some examples: "the heartbreak grape," "divine," "enigmatic," "fickle," "the Holy Grail of winemaking," "lightweight," "boring," and "expressive". Pinot Noir has garnered such divergent commentary for two reasons. First, it is a difficult grape to grow; it buds quite early in the season, making it susceptible to early frosts. This can limit its production, as can the fact that it's often attacked by mildew and other vineyard diseases. The grape bunches are compact and tightly formed, which allows for little air circulation and encourages interior rot. All of these factors culminate in low yields at harvest time. These issues are significant in and of themselves, before an ounce of wine is even produced.

The second major reason for the negative press is that even when the harvest is good, the wine produced can be ho-hum and uninspiring. In fact, it is often downright boring. This second issue is where the phrase "the Holy Grail of winemaking" originates. Despite the best efforts of winemakers all over the world, Pinot Noir rarely achieves greatness. Yet a good example of the wine will smell of tender fruits like cherries and raspberries with hints of flowers, such as violets and roses. Throw in some cedar, chocolate tones, and earthy smells like truffles and rotting vegetables, and the overall strange and complex mixture of smells is always memorable. The wine's taste, at its best, is very delicate, smooth, and velvety, providing lingering aftertastes that replay many of those original smells. The good wines from the Pinot Noir grape are enigmatic, seductive medleys prized all over the world yet seldom duplicated. This doesn't stop people from trying to make wine from this grape (and hearts from being broken due to failed efforts).

I must admit that, although I have tasted many lousy bottles of Pinot Noir, this grape is my favourite. I love the wine with a simple meal and often enjoy it without accompanying food. Compared to Cabernet Sauvignon, which offers great quality and consistency, Pinot Noir, when great (which isn't often), achieves greater heights. Part of its allure is probably the fact that it *is* enigmatic and fickle. When you're lucky enough to get one that's on character, the pleasure is amplified.

I'd be remiss if I didn't mention that wine experts, even Jancis Robinson, describe Pinot Noir as "feminine." The reason is not its sometimes enigmatic qualities but rather its delicate texture, integrated tannic structure, and intriguing combination of smells. The Cabernet Sauvignon, by contrast, seems bolder, harsher, and brasher, with clearly defined flavours — more male perhaps? Anyway, the comparison has some merit and is a useful way to differentiate between the wines produced by these classic grapes.

Pinot Noir and Champagne: Pinot Noir and the lesser-known Pinot Meunier are two red grapes used in many Champagne blends. When combined with Chardonnay, Pinot Noir adds complexity to this famous bubbly (It also improves its prospects for aging.) The higher the proportion of Pinot Noir, the more pronounced the flavours — and the tastier the wine, in my opinion.

Historical Benchmark: The Pinot Noir is one of the oldest known grape varieties. Since the Middle Ages, it has had a place of honour in the Burgundy area of France (and in Champagne as a blending grape). At its peak quality, the grape makes a strangely subtle yet expressive wine. Great Burgundies will age well, holding a nice acidity and developing bouquets that can only be described as ethereal. It's a seductive wine, often overpriced but well worth experiencing. Only in Oregon have other winemakers begun to consistently produce great Pinot Noirs — a bit of a surprise for some Burgundians! Other areas are also showing promise: New Zealand and Ontario, Canada in particular.

PINOT NOIR FACTS

Colour:	Black
Grown:	France, Germany, around the world (thrives in cool climates)
At Home:	Burgundy, Champagne, Oregon, Northern California, New Zealand, Ontario
Soil:	Clay, chalky areas
Character:	Lighter colour, delicate, complex aromas, cherry, raspberry, floral, cedar, chocolate, rotten vegetables and mushroom smells, velvety, subtle

(iii) Merlot

The Merlot grape has suffered a somewhat similar fate as the Chardonnay. It became the go-to red grape for many cocktail hour offerings; you'd consistently be offered a glass of Merlot or Chardonnay. In the famous wine movie *Sideways,* the lead character pleads, "[Give me] anything but Merlot." He was fed up with Merlot's too-frequent appearances at parties and as a restaurant house wine, often in one-dimensional incarnations.

Traditionally, the Merlot was used as a Bordeaux blending grape in concert with the Cabernet Sauvignon and Cabernet Franc. Soon it became fashionable as a standalone varietal grape. It makes smooth, easy-drinking red wine — very approachable because the tannins are quite soft and the grape is generally low in acid. The aromas are very pleasant, showing hints of chocolate, plum, strawberry, black cherry, some floral notes, and background spiciness. These attributes are all wrapped in a velvety smooth mouthfeel. No wonder Merlot became so popular among mainstream wine drinkers. Of course, the problem was that a lot of mediocre Merlot was produced to fill the demand. Soon it became *vino non grata.*

The original greatness of this classic grape was its ability to blend with other grapes, especially the Cabernet Sauvignon. The latter grape shows great structure and flavour characteristics, which sometimes are almost too defined, almost awkward — like a bull in a china shop. Then Merlot is blended in, and it fills in the spaces. If you imagine a North American native teepee, the Cabernet Sauvignon represents the wood pieces that create the tent shape, and the Merlot is the collection of skins that complete the structure. The blend is an ideal partnership: a wine with great complexity and finesse. In California, you'll often see the term *meritage* to denote the combination of the Cabernet Sauvignon, Merlot, and often Cabernet Franc grapes (sometimes the Petit Verdot and Malbec grapes are also included in the blend). This is the New World version of red Bordeaux wine.

Historical Benchmark: Today Merlot is the most widely planted quality grape in France. Its claim to greatness originally came from its role as a blending grape. Used in conjunction with Cabernet Sauvignon, Merlot achieves tremendous results. It is now commonly seen as a standalone varietal, where it can shine as a fine wine[45] . . . but look out for the cheap, poorly made versions, as they spoil Merlot's reputation.

MERLOT FACTS

Colour:	Black, blue tints
Grown:	Worldwide (thrives in moderate, warm climates)
At Home:	France (Bordeaux and southern regions), California, Australia, Chile, Canada (British Columbia)
Soil:	Clay
Character:	Fruity, chocolate, strawberry, black cherry, spicy, plum, smooth, full tasting

(iv) Syrah/Shiraz

This grape is the only classic grape that goes by two common names: Syrah and Shiraz. Each name, to some extent, indicates a certain style of wine. The Syrah is often a more restrained yet complex wine, whereas the Shiraz, popularized in Australia, is more fruit-forward and packed full of flavour. Despite these differences in style, this grape is known for producing *powerful* wine.

The overall confusion of the grape's "real" name dates back to ancient times. While the grape was traditionally used in Southern France, some people speculate that the grape originated in Iran (ancient Persia) near a town by the name of *Shiraz*. One legend claims that the Phoenicians brought the grape to the Rhône area. Another story has the grape arriving with a returning French crusader a thousand years ago. Others claim the grape is a result of crossing local varieties. Whatever the truth, the origin of the name

....

45   *Château Pétrus:* In the Bordeaux region of Pomerol, the Merlot is used in higher proportions in blends, often exceeding the Cabernet Sauvignon. In some cases, 100% Merlot wines do achieve great quality. At the famous Château Pétrus, the wine is made almost exclusively from Merlot. This wine is often the world's most expensive red wine, costing well over $1,000 a bottle.

makes for great debate and is a good side note to all things relating to stories about wine.

The Syrah is similar to the Cabernet Sauvignon in two main ways. First, both make strong, bold wines. Second, historically, each grape earned its reputation as a key grape in blended wines. The Syrah was mainly blended in Southern France, most notably in Châteauneuf-du-Pape. It was also frequently added to the wines of Bordeaux in the 18th and 19th centuries, prior to strict rules disallowing those practices.[46] The Syrah is a trusted addition to blends due to its ability to produce powerful flavours. However, it also gained fame as a varietal in the northern Rhône region of France.

The small regions around Hermitage in the Rhône Valley (including Cornas, Saint-Joseph, and Côte-Rôtie) are where Syrah shines as a single-grape wine. These wines can be quite similar to great Cabernets and will age for years. The other region of the world where the Syrah, now called the *Shiraz*, is commonly used by itself is South Australia. In fact, Shiraz is the most widely planted grape in the Land of Oz. Following the French precedent, a usual combination of varieties blended by the Australians is Shiraz–Cabernet Sauvignon, which surprisingly works.

Historical Benchmark: The Syrah grape's traditional home is southern France. There, you'll find very dry, dark wines with aromas of cassis (black currant), tar, dark fruit accented by mint, and a hint of smokiness (some smell like wet tobacco). The grape also produces very tannic wines, which helps explain its aging potential. The Syrah's suitability to oak barrel-aging is well known, and this adds that pleasant vanilla, woody bouquet you'll find in older wines.

SYRAH/SHIRAZ FACTS

Colour: Black

Grown: South France, Australia, California, Chile, South Africa

. . . .

46   *Claret:* The English referred to red wines from Bordeaux as *claret* (*Clairet* in French) since the 12th century, when England owned this part of France and called it Aquitaine. The wines were lighter in colour in those days, and Syrah wines added colour and flavour.

At Home:	Rhône Valley, South Australia
Soil:	Rocky, poor soils
Character:	Tannic, dark colour, vanilla, tar, cassis, tobacco/smoky, spicy dark fruit, mint, licorice

(v) Other Important Red Grapes

The classic red grape varieties we've just discussed make up the majority of great red wines — probably over 75%. These grapes make complex and interesting wines in the hands of good grape growers and winemakers. However, some other red grapes also produce fascinating wine.

NEBBIOLO

In the foothills of the Alps in northern Italy grows a grape that produces some tremendously age-worthy wines. In fact, these wines are almost *undrinkable* during the first five years of their life in the bottle. When young, they represent the ultimate example of a *closed wine* — a wine that won't open up and disclose its hidden smells and flavours until it goes through adequate aging.

Nebbiolo-based wines yield a traditional tar and roses nose that is their signature aroma. As the wine matures, other smells emerge, like violets, smoke, licorice, leather, and vanilla (when the wine is young, the aroma will include cherry and anise). The grape itself is black, but it does not impart great colour in wine. The wine is a lighter red at the beginning and turns an orange/brick-red hue when older. Besides of its interesting smells and colour, the wine is always quite acidic and tannic (in this case, the tannins are responsible for the evolution of the bouquet; they change their chemistry over the years and combine with the other elements in the wine). When aged sufficiently, Nebbiolo produces a full tasting sensation and complex textures on the palate. It is a BIG wine!

Traditionally, the Nebbiolo has excelled around the villages of Barolo and Barbaresco. Italians have been making wine from this grape since at least the 14th century, and the winemaking method changed little until the

1970s. After fermentation, the wine was stored in huge oak barrels for a year or more. Once bottled, it still required significant aging to soften and evolve into a pleasant-tasting wine. There was a certain roughness about the finished product that was quite alluring. All the strange smells accompanied by the wine's chewy texture created a kind of cult following of the Barolos in particular. Since the 1970s, modern winemaking techniques and small French oak barrels have created a smoother, fruitier wine. It also started a war of ideas. On the one side were the traditionalists, championing the amazing age-worthy wines of old; on the other side was the younger generation that appreciated the fresher, livelier, and more approachable tasting wines. Who's right? If you want to drink a Barolo in two to three years, go with the kids (new techniques). If you can store wines for ten to twenty years, get the traditionally produced variety.

Nebbiolo is not a grape that travels well (just like the Pinot Noir). It thrives in the strong, limestone-rich, muddy soils found in northwest Italy in the Piemonte region. You can find plantings of this grape around the world, but rarely do they produce good wine. There has been limited success in Australia, Argentina, South Africa, and the United States, especially in California and Oregon. Overall, wines from the Nebbiolo grape are rare, interesting, and worth tasting.

SANGIOVESE

Italy is also the home of another great red grape used in Chianti, the famous wine from the Tuscany region. The grape, Sangiovese, is the most widely planted red variety in all of Italy. In the past, it was most commonly used in blends.[47] Only in Tuscany is the Sangiovese the predominant, or only, grape in a red wine. The most prestigious, and arguably the best, Sangiovese-based wines come from Brunello di Montalcino, a small area within the Chianti region. These great wines are made with 100% Sangiovese.

The wines from the Sangiovese, to my palate, can be quite similar to Bordeaux from France. They have a nice balance of acidity and moderate to high levels of tannins. The nose can reveal concentrated smells of flowers,

47   *Chianti:* Historically, Chiantis were made with over ten approved grape varieties, including two white grapes!

cherries, and strawberries when the wine is young and a certain oakiness, nuttiness, and spiciness as it ages. This wine can be enjoyed when it is both young and old.

Outside of Italy, the grape is gaining popularity. In California and Australia, the wines tend to highlight the fruitiness of the grape at the expense of subtlety. Too much sunlight and warmth is not necessarily a benefit to Sangiovese-based wines. Some more typical (i.e., Chianti-like) wines are now being made in Washington State and Oregon. The other area that is making some notable wine from this grape is Western Australia. In the cooler regions like Margaret River, the Aussies are making inroads towards quality Sangiovese wine; it is becoming a popular choice among imbibers from Down Under. Fair dinkum!

GAMAY

This grape was once a dominant variety in Burgundy. It was so popular in the 14th century that it was eventually banned in all parts of the region except Beaujolais because it was displacing the "more-refined" Pinot Noir. The main reasons for the Gamay's abundant planting were its ability to ripen early in the season and its consistently generous yields. Unfortunately, it did not produce great wine — good wine but not great.

Today the grape is most famous for its easy-drinking qualities. The wine is invariably full of fruit flavours and aromas of fresh cherries and strawberries. It is always low in tannins, which contributes to its smoothness. I consider this wine to be a terrific quaffing wine; serve it along with a light white wine like a better Pinot Grigio.

Beaujolias Nouveau:[48] There is a fun tradition in Beaujolais that is celebrated on the third Thursday of every November. On that date, the current year's vintage is allowed to be released — in other words, it is an extremely young wine! There is no aging. Trucks are loaded up in Beaujolais, and at

....

48 *Carbonic Maceration*: Carbonic maceration is a technique used for Beaujolais Nouveau where the grapes are not crushed immediately and the juice ferments inside the skins. This creates a low-tannin, easy-drinking wine that you can drink when it's only a few weeks old.

midnight of that third Thursday, people rush to the bistros all over France; the wine is also shipped around the world by airplane. I was once in Paris the day the *Nouveau* (new) wine was poured for the first time. The atmosphere was loud and joyous, celebrating the first wine of that year's vintage. The Roman god of wine, Bacchus, would have been pleased.

Beaujolais Cru: The region of Beaujolais isn't all about quaffing fruity wine. Ten small subregions of Beaujolais make serious wines, known as *cru* wines. They follow more-traditional winemaking techniques and often use wood barrels for aging prior to the bottling. My two favorites are called Fleurie and Moulin-à-Vent. We'll discuss the whole concept of the term *cru* in Section Three.

TEMPRANILLO

Now we turn to Spain for the final important red grape variety. Tempranillo is an interesting grape that produces relatively delicate red wines despite being grown in a hot climate. Usually, hot regions produce almost burnt tastes and unbalanced touches of tar and tobacco smells in their wines. Not so with the Tempranillo, which is most commonly used in Spain.

The Tempranillo is often used in blends in the Iberian Peninsula (Portugal and Spain). Its greatest expression comes from the Rioja region in north central Spain. Rioja is a blend of the Tempranillo (around 60% typically) and a few other grapes, of which the Granacha[49] is the most important. Riojas are usually aged in American oak barrels, and the best-made wines are known to age very well. Over the last few years, winemaking practices in Rioja have evolved. The wine has gone from a bold, almost astringent style to a more accessible, fruitier and delicate wine. I miss the old stuff!

Typically, wine from the Tempranillo grape shows a nice ruby colour when young. The colour eventually turns to a brick-red/brown in aged Rioja. In the young wines, the nose displays cherries along with strong vanilla aromas

....

49   *Grapes names:* One problem with keeping track of the names of grapes is that many have different names in different countries (sometimes in different regions within the same country!). The Garnacha is *Grenache* in France; that's easy, but the grape also goes by another twenty or more names all over Europe.

(this attribute was more pronounced in the older style when the wines were commonly aged in oak for up to five years). As the wine ages, people also note smells of nuts and caramels, along with dark fruit and sometimes wet tobacco. Overall, the well-aged examples of Tempranillo offer an interesting — and for me, addictive — combination of flavours, mouthfeel, and bouquet. Today, the Tempranillo is also grown in California, in Argentina, and sparsely in other parts of the world. Some think Tepranillo may become an important single-varietal wine. As usual, the trick will be finding suitable soils and climate. With further experimentation and planting, it may indeed be poised to become a more popular wine around the world.

IV. FORGOTTEN GRAPES

*"Alone in the vegetable kingdom, the vine makes the
true savour of the earth intelligible to man."* — Collette,
French novelist and poet, "Prisons et Paradis"

As I mentioned earlier in this book, perhaps some of the classic and other important grapes became great due to good luck or accidents of history. A grape like the Merlot may have become popular because of the relative ease in growing the vine itself and its consistent outcome in winemaking. Any grape would have been welcomed by an early winery if it yielded productive harvests and balanced, full-tasting wines. Yet the Merlot just as easily could have been passed over for another grape hundreds of years ago. Nevertheless, since that time a lot of people have grown Merlot, nurtured it, found ideal soils, and perfected it as a wine, and the rest is history, as they say. Was Merlot always a great grape for winemaking, or was it nurtured, bred, and developed into a classic variety? A provocative question, in my view.

There are a number of grapes for which there is some consensus regarding their potential worth as quality wine grapes. I'll discuss a few of these notables (no doubt my choices, and omissions, will create some controversy!).

OTHER NOTABLE WHITE GRAPES

Furmint: A grape used in the famous sweet wines from Tokaj, Hungary, known as Tokaji Aszú. These wines are dark yellow and full-bodied, and they have a nice, creamy mouthfeel. They are often compared to Sauternes from Bordeaux.

Müller-Thurgau: A cross between the Riesling and Sylvaner grape, Müller-Thurgau produces an easy-drinking wine. Its character is best described as "refreshing." The nose is quite floral, and the wine is lightly acidic on the palate with a nice, clean finish. The grape is a varietal workhorse in Germany, being the most planted variety; it is early-ripening, vigorous, resistant to cold, and comfortable in many soil types. No wonder grape growers love it!

Marsanne: The Marsanne grape is famous for making white wines of some depth in the northern Rhône region of France. The wines are labeled Hermitage, Crozes-Hermitage, and Saint-Joseph (remember, the French normally use place names versus grape names on their labels). The wines from the Marsanne grape, which are noted for their longevity, are a favourite of the great wine critic Hugh Johnson.

Palomino: The Palomino is used mainly in Sherry, Spain's most famous wine. It is especially suited to making this fortified wine (any style). In my opinion, Palomino deserves to be included — for this reason alone — as a notable variety.

Pinot Blanc: This grape can produce some distinctive wine, especially when given some barrel-aging. It is sometimes compared to Chardonnay but generally lacks its complexity. However, Pinot Blanc makes a nice food wine because its naturally high acidity cuts through rich flavours and sauces. The Pinot Blanc is also often used in sparkling wines because of its acidic qualities.

Verdicchio: Used mainly in the Marches region of Italy, this grape creates a straw-coloured wine and often tastes of almonds. It is high in acid, which makes it a good blending partner in sparkling wines. It is not grown outside of this area, and as a result, its full potential is unknown.

Welschriesling: This grape, which originated in central Europe, produces a delightful flowery-smelling and fruity wine. The fact that the grape's name includes *Riesling* causes some confusion (in Germany, they'd prefer that the name would be changed completely). However, the Welschriesling's flavours *are* reminiscent of the Riesling. It is grown in parts of Italy, Romania, Hungary, and Austria, where it is used not only in table wines but also in sweet, late-harvest wines.

OTHER NOTABLE RED GRAPES

Barbera: This grape is one of the most commonly grown red varieties in Italy. Good examples of the wine, usually from moderate climates, have a nice structure due to naturally high acidity and tannins (making these wines a little rough when young) along with pleasant fruit flavours. The Barbera has good aging potential, and some feel it evolves into a fine, if not great, wine. The grape is also widely used in California.

Grenache: The Grenache grape (known as *Grenacha* in Spanish) is associated with southern France and Spain, where it is thought to have originated. By itself, Grenache makes an easy-drinking, openly fruity wine. Part of its fame rests on its ability to make fine rosé wines. The rosés from Tavel, France, are a personal favourite. The Grenache really shines as a blending grape; it's a component of many wines for which "the sum is greater than the parts." The best blended wines with Grenache in the mix are Châteauneuf-du-Pape and many Riojas.[50] My particular affection goes to the now-trendy GSM blends (Grenache, Syrah, and Mourvèdre grapes) from Australia — wines with power, complexity, and, I suspect, great aging potential.

Malbec: This grape has long been a blending grape in Bordeaux, where it is most known and useful for its colour. Malbec's original fame as a varietal came from the "black" wines made in France's Cahors region. The grape is thick-skinned and yields very dark-coloured wine. The grape has experienced a renaissance of sorts with the rise in the popularity of Argentinian Malbecs. These wines are of good value and considerable weight (i.e., big, bold, velvety, and high-alcohol wines). I find that some of the cheaper examples of Malbec are too singular in their spicy, salty, tar taste and can be quite ghastly! However, good Malbecs make excellent food companions.

Petite Sirah: A little confusing, as this grape is correctly named Durif and is unrelated to the Syrah. The confusion comes from some of the grapes' similarities. The Petite Sirah is one of my favourites, probably due to my having tasted some very old Californian Petite Sirahs very early in my wine education. The wines can smell of spices, especially black pepper, and they have

---

50   *Grenacha in Spain:* The grape is a major component in one of the world's greatest red wines, the complex and silky Vega-Sicilia.

great tannic structure that softens with age as the complexity of bouquet and taste increases.

Touriga Nacional: At last we get to Europe's often forgotten, southwestern country of Portugal (except when it comes to its fortified Port wines). The Touriga Nacional is an important grape in Port, but it can also shine as a dry red wine. Its berries are quite small, and the fruit produces very dark wine with terrific aromas and considerable complexity. I like it because it's so different from the run-of-the-mill Cabernet- and Shiraz-based "value" wines that have flooded the market. Try buying a red Dão (named for a small region of Portugal) next time you visit your local wine shop; I think you'll be happy with your purchase and intrigued.

Zinfandel (Primitivo): For many years, people thought the Zinfandel grape was a native of California. However, it is now considered to be a descendent of the Italian Primitivo. It makes quality wines when cared for properly. Unfortunately, it never gained widespread popularity, partly because consumers can rarely tell what kind of wine they're getting; the grape is malleable, easy to grow and capable of making many styles, from pink rosés to light, fruity reds to full-bodied, high-alcohol reds. The best examples of Californian Zinfandel or Italian Primitivo make a raisiny, fruit-forward (think raspberries and blackberries) wine that settles into a subtler and more enjoyable drink over time.

IGNORED OR LOST GRAPES?

My instincts tell me that there are probably a good number of virtually unknown grapes in the world that could produce fine wine — maybe great wine. Recently, I tried two wines, a white and red, from the small Bierzo region in northwestern Spain that were absolutely delightful. The white, which was made with the Godello[51] grape, showed aromas of grapefruit, melons, and peaches with a nice background minerality. The taste displayed absolute freshness with lingering, subtle hints of apple and spice. The Menica grape was the basis for the red wine. It reminded me of a lighter

51 *Álvarez de Toledo:* The Godello grape used in this wine produces a beautiful, flowery wine with considerable depth. Even Robert Parker gave it 90 points on his 100-point scale!

Pinot Noir wine. This wine offered raspberry, mineral, violet, and spicy aromas along with a medium body, smooth mouthfeel, and lively acidity. Both of these wines were real finds for my collection. The experience reminded me of so many other times I'd discovered obscure grape varieties or radically different approaches to a well-known grape (e.g., using the Cabernet Franc in ice wine — amazing!).

These two Spanish grapes are only a drop in the bucket of ignored or lost grapes. It's amazing to realize that Italy alone has over 300 grape varieties! Greece has a number of unique grape varieties that have been used since ancient times and that are producing some pretty good wines — grapes like the white varieties Moschofilero and Assyrtiko and red ones like Agiorgitiko and Xinomavro. Yet they have not been grown seriously in other regions. I wonder what might have become of these grapes if they had been grown in other climates and soils and if they'd been given the attention and pursuit for quality that was given to the classic grapes we now enjoy.

And what about all the other lost or ignored grapes from the Middle East and areas around the Black Sea, where grape cultivation and winemaking had its origin? How many grapes must have missed the caravan to the emerging wine-producing powerhouses in Greece and Rome and, by extension, the whole of modern Europe? Since most of the Middle East turned to Islam, a religion that prohibits drinking alcohol, the fate of many of those ancient grape varieties — and any vineyard remnants — has faded from memory, most likely torn from the ground well over a thousand years ago.

I fantasize that we are in store for a few discoveries yet. Maybe a few vines of a long-forgotten grape are still alive in some small village in Armenia. An adventurous vigneron may find them and bring them back to California. After a few decades of careful breeding and experimentation with different sites, a great wine might emerge from this ancient variety. A new classic! Who knows?

CONCLUSION TO SECTION TWO: THE GRAPES

This section may have seemed a little overwhelming, because it's filled with nomenclature, geography, and history. But memorizing a few grapes is essential to getting comfortable in the world of wines. I've tried to keep the discussion to the main facts and added some humour and anecdotes to aid your memory.

Step two is remembering some of the grapes' common, predominant characteristics. Sound daunting? It isn't. Reviewing my descriptions as you drink various wines will help. Over time, you'll be surprised how easy it is to recall grape characteristics.

In reality, you really only need to know the eight classic grapes and the secondary nine important grapes — seventeen in all:[52]

See next page for a table summarizing all the important grapes.

---

52   *Classic versus Important:* Some people will undoubtedly disagree with my classifications. What I don't deny is that someday, my Important Grapes could easily become Classic (and vice versa!).

	Red	**White**
Classic	Cabernet Sauvignon	Chardonnay
	Pinot Noir	Riesling
	Merlot	Sauvignon Blanc
	Syrah (Shiraz)	Gewürztraminer
Important	Nebbiolo	Chenin Blanc
	Sangiovese	Muscat
	Gamay	Pinot Gris (Grigio)
	Tempranillo	Sémillon
		Viognier

The big seventeen collectively make up over 90% of the world's greatest wines. By knowing these varieties, you'll be well-versed in quality wines. This knowledge alone will give you a firm footing in the wonderful world of oenology.

SECTION THREE: THE APPROACH

I. THE CHOICES

"The juice of the grape is liquid quintessence of concentrated sunbeams." —Thomas Love Peacock, Melincourt

A winemaker has to make countless decisions related to growing grapes, making the wine, and then marketing the end product. Yet fundamentally, every decision will flow from two critical and central decisions. First, the winemaker must decide whether to use just one grape (and produce a wine we call a *varietal*) or to use two or more grapes and create a blend (or *cuvée* in French). It may surprise you that the majority of the wines from the traditional European winemaking countries are actually blends.

The second critical choice is one of quality. Does the winery want to make a value-based wine attractive to the average consumer? Or will they make a high-quality product for only a limited audience (i.e., experts and/or wealthy customers). Wine quality runs a huge spectrum, from the quaffing bulk wines, or *jug wines,* to the exalted (and expensive) transcendent wines. Thousands of wines find their place between those extremes.

Of course, some critics note that many winemakers have no choice whatsoever regarding grapes or quality. This is correct. A winemaker in a fine Bordeaux château is already committed — actually, *required* — to blend certain grape varieties and to do everything he or she can to make a high-end product. However, at some point in time, those central decisions had to have been made.

Next we will look briefly at the development of two regions in France, Bordeaux and Burgundy, and explore their history in winemaking. I will speculate about why they made their respective decisions regarding the blending, or not, of grape varieties. I'll follow with a discussion of the choices winemakers around the world must make to produce either value wines or quality wines. You'll learn about the consequent practices they've adopted in the winemaking process and explore the reasoning behind them.

Finally, we'll look at the clues you can find on labels, which will give you an indication of the approach the winemaker has taken regarding the grapes and the quality. An emerging, almost universal logic governs wine label content. You'll look at how and why that logic came about and learn the key words and phrases associated with the main rules and regulations.

II. TO BLEND OR NOT TO BLEND

"...the pure Septembral juice..." Francois Rabelais,
The Life of Gargantua and of Pantagruel

Many novice wine drinkers are greatly confused by wine labels that have no grape names. Consumers have become used to the ubiquitous appearance of grape names like Chardonnay or Cabernet Sauvignon on bottles of wine. Yet this practice was not very common in the world of wine until recent times.[53] Using grape names on labels became popular when California became a wine world powerhouse. Wineries in that state, especially starting in the 1970s, labeled their wines predominately by grape type (versus a place name). They also favoured single variety wines, known as *varietals,* which then became the norm in many emerging wine countries like Australia, Chile, and New Zealand. As a result, today's North American consumer, when presented with a European wine, is often confused about its contents because there is no grape name, just a place name like Vouvray.

A pivotal question is whether a blended wine is better than a varietal. This debate has gone on for centuries and will go on for many more. Those who champion blending argue that good blended wines are often greater than the sum of their parts. The end-product exceeds what each individual grape can achieve going solo in a wine. However, defenders of the varietal approach argue that the varietal wines showcase the beauty of one grape variety — simple, pure, and unadulterated. They also argue that varietals more clearly express the influence of the soil. It is simplicity versus complexity.

Let's look at the two famous homes, respectively, of each approach: blending and not blending.

....

53   *Grape names:* Only in certain parts of the Old World, mainly Germany and Alsace, were grape names commonly found on labels.

(i) The Cuvée

The French word for blend is *cuvée*. In the winery, this term can refer to any number of ways to blend: the blending of wines from different years, from different vineyards or regions, from different barrels or vats, or from different grapes. The last definition is the one important for our purposes: the blending of two or more grape varieties. And the most famous place in the wine world where this practice occurs is in the western region of France, Bordeaux.

THE STORY OF BORDEAUX

> *"He who aspires to be a serious wine drinker must drink claret"*
> —Samuel Johnson, *Boswell: Life of Johnson*

Bordeaux, the fourth largest city in France, is a very important port. It has been a major trading city since Roman times. This commercial character is very important to Bordeaux's history and to the development of its wine trade. Bordeaux is blessed with a large river, the Gironde, which runs to the Atlantic Ocean; this waterway was the city's main access to foreign trade. Moving inland, the river splits into the Garonne, running southeast, and the Dordogne, moving eastward. Interspersed among these three rivers are some of the greatest vineyards in the world.

The Romans called the area Aquitania, meaning "land of waters," because of the three rivers. In medieval times, it was controlled by the English and known as Aquitaine. Soon, the wines of Bordeaux became favourites of the English. The most important wines, the reds, were named *claret* due to their light colour (quite different from today's claret, which is brick/dark red and often full-bodied). As the centuries passed, Bordeaux became an important centre of trade. Merchants from all over the world set up shop in the port. Great buildings and, of course, warehouses were built throughout the city. Commerce, not the Church, was the predominant influence in this region, which may well explain why Bordeaux escaped some of the radical changes wrought by the French Revolution. Somehow, during these turbulent times, the châteaux remained mostly untouched and continued to operate as intact, complete estates right up to modern times.[54]

At some point in Bordeaux history, a number of grape varieties gained popularity and eventually were blended together. The most important of these grapes were the Cabernet Sauvignon, Merlot, Sauvignon Blanc, and Sémillon. My theory is that because Bordeaux was a centre of trade, many different wines, and eventually vines, travelled through this prominent port. Some would have come from inland via the Dordogne and Garonne rivers. Other wines and vines would have arrived from other regions by way of the sea. Either way, the winemakers of Bordeaux had access to many grape varieties and wine styles. Perhaps having tasted various wines from other parts of the world, they decided to continually plant new varieties and experiment with different combinations. As both typical business folks and winemakers, they could have simply been trying to improve their product and make it better and therefore more attractive and saleable! Anyway, at some point, winemakers in Bordeaux decided to always blend grape varieties. The question is why.

PLAYING IT SAFE

The great advantage of blending different grapes is that it gives the winemaker a certain degree of insurance (and some would argue, a better wine). The reason for this is that grapes often mature at different times of the year. Some do well in hot weather, while others prefer cooler, more moderate growing seasons. At harvest time, the winemaker can determine which grapes should make up the cuvée, or blend, according to their ripeness. If the Cabernet Sauvignon did not ripen completely and is quite acidic, then he or she can add extra ripe grapes like the Merlot to compensate. The

---

54   *The French Revolution:* The two main targets of the revolutionaries were the Church and the kings, along with all other nobility of that period (1789–1814).

winemaker in the cuvée blending tradition becomes more like a *chef*[55] in the kitchen. They both mix their ingredients to create a final product that is much different, and sometimes better, than its component parts.

Once more, perhaps, the commercial nature of the Bordeaux wine merchants steered them towards this blending mentality. Year in, year out, they could have a better guarantee of quality by blending rather than relying on just one grape variety. A single grape may or may not ripen on time and sometimes yielded a poor harvest. The winery must then deal with limited amounts of grapes some years with varying levels of quality — not a scenario that astute businessmen relish! Whatever the reason (and I've not seen any definitive explanation), in Bordeaux you'll find reds with as many as five blended grapes: Cabernet Sauvignon, Merlot, Cabernet Franc, Malbec, and Petit Verdot.[56] The whites will be blends of the Sauvignon Blanc and Sémillon (some Muscadelle is rarely used). Blending grapes is not only the story of Bordeaux; many other regions around the world followed the same practice. Blending became the new normal (tradition).

TWO BLENDED RED WINE POSTER BOYS

What are the limits of this blending? That is, how many grapes should be used? There's no correct answer to that question, but let's look at two interesting poster boys for red wines.

A fiasco or not? Italy's best known wine is Chianti, which is produced in the region of Tuscany. Traditionally, this wine was stored in a bottle known as a *fiasco* — a bottle with a thin neck and a bulb-shaped bottom covered in straw. The purpose of the straw was probably to protect the bottle during shipping. The name *fiasco* came from the Italian phrase *fare fiasco,* meaning

....

55   *Chef de Cave:* This term is most commonly used in Champagne, where blending is very complex and a real art form; the winemaker needs to maintain a consistent house style each year using wines from various growers. *Chef de Cave* — or *de chai,* as they call aboveground storage areas in France — is an apt term for all winemakers who embrace the blending philosophy. They are sometimes referred to as *Masters of Assemblage* in English.

56   *The sixth:* The Carmenère grape is a sixth grape used in Bordeaux, but it's used very sparingly. This grape is widely used in Chile, where it was often mistaken for the Merlot grape.

"to make a bottle." How *fiasco* came to mean "screw up" in English is open to debate. Some people think it's because Chianti, for years, was often a thin, astringent red wine that was so terrible it could hardly be called *wine;* it was a fiasco. Probably not true, but it makes a good story.

The interesting thing about Chianti red wine is that there are thirteen approved grapes allowed into the mix, and two of these grapes are white. Until recently, many wineries added too many inferior grapes, including a big proportion of white grapes, resulting in an overpriced but rather neutral tasting wine. Thankfully, this practise has changed. Today, the predominant grape is the Sangiovese, which is blended with small amounts of other quality grapes like the Canaiolo, Merlot, and sometimes Cabernet Sauvignon. Chiantis of today are anything but a fiasco. Many are brilliant and often age-worthy.

The pope's home: Our other poster boy for the practice of blending grapes is found in the south of France along the Rhône River. In an area surrounding the small town of Châteauneuf-du-Pape evolved one of the great fullbodied wines of the world.

The wine was given the name of the town, which literally translates as the "new home of the pope." In the 14th century, there were at times two popes (and even an anti-pope). This fascinating chapter in the history of the papacy was full of intrigue. During this period, one of the two popes moved to France and set up shop in a small town that was appropriately named Châteauneuf-du-Pape.[57] Of course, he needed wine to celebrate the Mass!

The recipe for Châteauneuf-du-Pape wine permitted fourteen possible grape varieties. As with Chianti, white grapes were included in the list — but here, they allowed six! The other eight, which are obviously reds, include the Grenache, Syrah, and Mourvèdre; this trio of grapes, known simply by the acronym GSM, now makes up the heart of all great Châteauneufs and has been adopted in other areas around the world. As I mentioned previously, GSM is one of my favourite combinations.

....

57 *Pope Clement V:* Clement was the first of the popes to live in France. His successor was John XXII, who built a permanent residence among the vineyards.

Many developments in history are compared to a pendulum, events causing the fashions and politics to swing from one extreme to the other. The story of wine also benefits from such a comparison. As we've discussed, blending was common in some parts of the Old World (mainly European wines) as well as in ancient Mediterranean cultures. Since the 1970s, the varietal wine styles have dominated in many parts of the world, especially in North America. Having the name of the grapes on the label — preferably just one grape — was important. "Give me my Chardonnay!" and "I only drink Cabs!" were common declarations. But times are changing, and blends are making a comeback. One good example comes from the home of the varietal revolution.

Meritage: In California, the term *Meritage* is used to describe a combination or blend of traditional Bordeaux grapes in a wine.[58] Therefore, a Californian white Meritage is made with Sauvignon Blanc and Sémillon. A red includes two or more of the iconic Bordeaux red grapes: Cabernet Sauvignon and Merlot along with Cabernet Franc, Malbec, and Petit Verdot. I believe that this single development is symbolic of an overall trend towards more blended wines in the marketplace.

Blending philosophy: Back to the central question: Why blend grapes instead of using just one variety? We've already discussed some of the reasons in "The Story of Bordeaux." Remember the winemaker and wine merchant's quest for improvement. Secondly, by planting different grape varieties, the winemaker could *manage risk*; if one grape was a late ripener and the growing season was short, the winemaker simply increased the proportion of the early ripener to ensure higher sugar levels and maintain quality. In other years, if early ripening grapes were too ripe (lacking in acidity), the winemaker favoured a late ripener in the final blend.

A third reason has to do with human nature: Some people simply like to differentiate their wine and/or think they might be creating a better or new

....

58 *Meritage:* The Meritage trademark, which was adopted in the U.S. in 1988, indicates the wine is a blend of the grape varieties used in Bordeaux. The word comes from *merit* and *heritage.*

wine. They aren't satisfied with the status quo. Australian winemakers are notorious for their tendency to break or ignore the rules. Aussies blend grapes into untried combinations just for the sake of getting something different or unique.

Finally let's not forget that many winemakers work in wineries with established procedures or in regions with stringent rules. These winemakers couldn't change the grape varieties even if they wanted to mix up the blend, much less produce a varietal wine.

Other blending advantages: The proponents of blending grape varieties maintain that there are some other distinct advantages in the final product. Blended wines tend to be more complex because of the multiple characteristics each grape brings to the cuvée. For example, a dark, heavier-tasting red grape could be enhanced with a fruitier, fresher-tasting grape variety. This overall complexity yields more flavours and makes the wine more interesting to your palate. The greater complexity also enables you to pair the wines with more types of cuisine.

Perhaps the greatest argument for blending grape varieties is summed up in one word: *balance.* These wines tend to be more *well-rounded* than varietal wines. One of the best examples is the combination of the well-defined, structured Cabernet Sauvignon (with its ample tannins and distinct flavours) with the fruitier, softer Merlot (smooth texture with big fruit flavours). In Bordeaux, winemakers may also add a little Malbec for velvety texture, a little Cabernet Franc for a violet bouquet, and a touch of Petit Verdot for spiciness and colour.

The art and science of blending grapes is a complex undertaking similar to the development and execution of classic recipes in a fine restaurant. The contribution of each ingredient, or grape, must be carefully measured and monitored from start to finish. Proponents of blending think it not only yields better balance but also results in a wine that's much more than what any of the individual grapes could have offered alone. In short, blends bring more to the table.

(ii) The Varietal

The trend towards single-grape wines accelerated in the latter part of the 20th century. Because of the success of the California wine industry, a lot of people came to actually believe that most wines were made with just one grape. They were in some parts of the world — most notably in Burgundy, France.[59] Let's begin by looking at Burgundy as an example of this approach to winemaking and consider some of the reasons for and implications of this varietal philosophy.

THE STORY OF BURGUNDY

> *"The first duty of wine is to be red. . . . The second is to be*
> *Burgundy."* —Harry Waugh, English wine merchant

Burgundy is located in the mideastern part of France and is situated along the river Saône, which runs south into the much larger Rhône River; in turn, the Rhône empties into the Mediterranean. It is worth noting that virtually all wine-growing areas of ancient times were situated on river banks. The reasons were twofold. First, rivers were the easiest route for travel and allowed quick access to the sea or ocean for widespread transport and trade. Second, river banks were often part of a larger valley and so provided ideal slopes for the vineyards. A gentle hill allows good drainage, which is essential to growing better grapes. Fertile flat fields produce abundant grapes with very little flavour, while well-drained slopes force roots of the vine to go deeper for nutrients and, ultimately, yield grapes with greater flavour. A vine must struggle a little to actually develop the kind of grapes worthy of making fine wine. Ancient civilizations came to understand this principle very well, as many of the great ancient vineyards were located in areas where virtually no other crop would thrive.

Historians believe the Phoenicians introduced wine into the Burgundy region before 1000 BC. The Phoenicians traded wine for other goods from

---

59 *Germany:* Using one grape variety is common practice in German wine regions as well. I picked Burgundy due to the general public's greater awareness of the region and Burgundy's use of two well-known red and white classic grapes, the Pinot Noir and Chardonnay.

the local inhabitants. At some point, the locals would have started growing their own vines and making wine. Evidence suggests this was well underway by the time of Caesar. Within a few centuries, the Roman Catholic Church became a major influence in the development of Burgundy's wine industry. In fact, the Cistercian and Benedictine monasteries were central in the history of winemaking in the area and, for that matter, in the overall development of winemaking practices (some of which have barely changed in the last 500 years). Soon Burgundy's wines became widely known as the finest in France and a favourite of the Vatican.

The power of the Church grew steadily in France, and its influence was keenly felt in the courts of the nobility and various kings. The pope's favourite wine soon became the wine of choice in Paris and Versailles. It was soon fashionable for various nobles to purchase Chateaux in Burgundy; as a result, Burgundy became a region dominated by both the French nobility and the Roman Catholic Church. This situation became a powder-keg during the French Revolution, especially considering the words of Denis Diderot, a voice of inspiration for the revolution: "Men will never be free until the last king is strangled with the entrails of the last priest."

In contrast to the region of Burgundy, the owners of the great Châteaux in Bordeaux were not as closely associated with the Church, and luckily most were not viewed as part of the aristocracy. Although the region did suffer during the Revolution, with a few folks being sent to the infamous guillotine, the large estates in Bordeaux stayed intact and have remained so to this day. Not so in Burgundy. The wrath of revolution was unleashed in the region when the monarchy was displaced. The Church quickly packed up or sold up its possessions, as much as possible, and retreated from the region ahead of the rebellious masses. Some of the nobility got lucky and sold out too, while others suffered the same fate as Louis XVI and his wife Marie Antoinette.

The new authorities seized all the land and divided it among the peasants, the result of which was a deeply fragmented vineyard ownership. This was exacerbated by French inheritance laws that required all holdings to be split among the surviving heirs. In Burgundy today, very few vineyards have only one owner. Some famous sites are owned by many people with holdings as small as a few rows! The nuances of each parcel of land became more and more noticeable due to these small operations. In my opinion, the whole notion of terroir[60] came into prominence in Burgundy due to this extreme fragmentation and the many owner-farmers. A wine made from grapes only one hundred yards away was sometimes noticeably different. People were able to notice the distinct influence of a vineyard's site, or terroir, due to one central reason: In Burgundy, winemakers mostly used only one grape for reds and one for whites, the Pinot Noir and Chardonnay, respectively. With the key factor of grape variety held constant, it was easier to notice differences due to the growing sites.

Why Burgundy settled on using just two grapes for its fine wines is a bit of a mystery (we are only talking about the fine wines from Burgundy and excluding Beaujolais, where the reds are made from the Gamay grape). Some think that the two grapes were native to the region, while others believe they were brought in as wine cuttings by the Greeks or Romans. Whatever the truth, the Pinot Noir and the Chardonnay are two extraordinary grapes. First, the Pinot Noir, among all red grapes, seems the most capable of producing a *complete* wine. It has unique aromas reminiscent of barnyard rotting smells that are accompanied by a background fruitiness of black fruit and flowery bouquets — an indefinable, strangely appealing mixture of smells. Taste reveals complex flavours, balance, fresh acidity, and subtle tannins, and it all ends with a long-lasting finish. At its best, this unusual grape, almost impossible to describe, is beguiling!

Second, we find the finest of all white grapes, the Chardonnay — a grape that is completely malleable in a winemaker's hands and also one that travels so well to different regions in the world (unlike the Pinot Noir!). The Chardonnay shows beautiful smells of apple, nuts, vanilla, yogurt, minerals, and a certain flintiness along with tastes so complex and diverse for a white wine that it could often be confused for a red. This wine also matches well with most meats and many cheeses. A real wonder! Perhaps there is no mystery in Burgundy's reliance on two grapes for its fine wines after all;

....

60   *Terroir:* Remember, *terroir* means not only soil but also microclimates, slope of the hill, drainage, aspect, and even the very air.

with characteristics like these, why dilute or pollute your wine with other grape varieties?

Interestingly, the history of Burgundy included attempts to protect these two grapes because of their superior quality. For example, in the 14th century, the Duke of Burgundy, Philip the Bold, banned the Gamay grape.[61] Its wines were considered inferior to Pinot Noir and still are used sparingly in the region (with the exception of Beaujolais, where the Gamay is king). The Chardonnay was also enshrined in regulations as the grape of choice, with the inferior Aligoté grape used only in the outlying areas.

There are two more reasons Burgundy will probably never change from the varietal approach. Both can be attributed to outcomes of the Revolution. Since most growers owned only parcels of a vineyard, they were at the mercy of *négociants* (wine producers and/or merchants). These winemaking businessmen would buy the grapes and make the wine in their cellars far away from the vineyard (only recently did it become common for wine to be made and bottled in Burgundy vineyards/wineries). These négociants controlled the market for over two hundred years and were able to dictate which varieties they would buy. Since the Pinot Noir and Chardonnay were considered the best, this alone would have discouraged planting new grape varieties. Plus, the growers' holdings were so small that planting various grape types would have been impractical. Better to concentrate on your few vines of one grape type, perfect the fruit quality, and ensure a healthy, abundant harvest. This combination of factors — spectacularly characteristic grapes, edicts from nobility, the external control by the négociants, and small vineyard holdings — all played a part in the focus that Burgundy continues to have on production: to make only varietals.

FOCUSING ON NURTURING

Given everything we've discussed about Burgundy and the varietal approach, it becomes clear that the work of caring for the grapes is paramount in this model. The overriding modus operandi in regions that rely

....

61   *Obedience:* The Church dominated Burgundy, so the people were used to being obedient. Maybe this is another reason winemakers followed edicts and regulations that required using one grape for whites and one for reds.

on only one grape variety for their wine is *nurturing*. You cannot rely on another grape coming to the rescue. If the harvest is a disaster, the winery will suffer, either in terms of revenue or reputation. All of the vintner's energy must be focused on getting the grapes growing, keeping them healthy, and harvesting them at the right time. The actual winemaking is almost secondary, as the goal is to showcase the grape by bringing out the very best from the vineyard.

The emphasis on nurturing may also go a long way toward explaining the obsession with terroir in Burgundy. As mentioned, there are stories of vintners actually tasting the earth to determine its suitability for growing grapes. As the Church gained more and more vineyards in Burgundy, either as a result from the cooperation of powerful Dukes or from purchasing land, the monasteries became centres for learning about viticulture (growing grapes) and the art of vinification (making wines). Through years of experimentation and hard work, the monks refined their techniques. The ultimate outcome was to allow the vineyard sites to *express themselves* through the grapes and wine. The monks were able to bring the Chardonnay or Pinot Noir to the pinnacle of perfection. The wine from each separate vineyard[62] had its own distinct character and nuances. These differences were carefully noted, and a catalogue of the very best vineyards, or *cru*, eventually found its way into today's categories of first, or finest, growths (we'll discuss this classification system pioneered in France later). As a result of the evolution of winemaking in Burgundy (and Germany), terroir and site selection became a worldwide science — a focus that is critical when you rely on just one grape for your wine.

THE GOOD, THE BAD, AND THE UGLY

The varietal approach to winemaking has a certain beauty in its simplicity. There is a purity to relying on just one grape. For me, only three grapes consistently rise to greatness: the Pinot Noir, the Riesling, and the Chardonnay. Other grapes make remarkable wines, but as varietals, something is lacking.

....

62   *Cru*: This term literally means "growth," but in the wine world, it refers to a
     specific vineyard. Monks carefully determined the best sites and built stone walls
     around them.

For example, even the greatest Cabernet Sauvignon is often missing a certain roundness; such wines always seem a touch austere, lacking soft textures and not quite balanced (to my palate, anyway). That is why a California wine labeled a varietal Cabernet Sauvignon almost always has a little of another grape blended with it (5–10%). I prefer blended wines in general, as I believe most grapes do better as part of a greater whole, but the Pinot Noir, the Riesling, and the Chardonnay are grapes that, in themselves, can make complete wines. They're the exceptions that prove the rule.

However, the success of these three grapes has created some unfortunate consequences in winemaking circles. Also, the varietal approach requires a determined winery, sometimes with deep pockets, to persevere with the one-grape-per-wine philosophy; this allows the winery to withstand the economic impact of poor harvests.

The Good: Nothing compares with a great red Burgundy. The Pinot Noir grape can, and I emphasize *can,* make a beverage that is virtually ethereal in its colour, smell, and taste. I've never forgotten my experiences drinking these wines in Burgundy. One memory especially stands out. My wife and I were traveling with our two children, and we were in a large cellar in Beaune. I had just finished tasting over a dozen reds (only a few ounces in total) and remember looking back at my son, 12 years old at the time. He had just sipped on a well-aged wine and had an almost bemused, satisfied smile on his face — an expression of both surprise and wonder. To me, this was a superb summarization of the magical character of great Pinot Noirs: delicious, mysterious, and seductive. I'm able to say as many good words about the Riesling grape as well. Shifting over to Germanic soil, and Alsatian too, you will find amazing wines made from just this one grape.

Finally, the Chardonnay rounds out my trio of great wine varietals. Here's a grape that is so malleable that it can go from making lean, minerally, muted, slightly fruity tasting wines in the north of Burgundy (and in cool climates almost everywhere) to yielding complex, powerful, multi-flavoured wines in moderate warmer climates.

In summary, these three grapes — Pinot Noir, Riesling, and Chardonnay — are truly remarkable. I've waxed on about each one of them as wines for pure enjoyment when drinking them on their own. They are also wonderful accompaniments to a variety of foods.[63] Their complex characters allow them to stand up to many different flavours, either as a harmonizing match for food or as a perfect foil or contrast that enhances food. It's also worth noting that these grapes make wines that are all age-worthy. Three interesting grapes to celebrate!

The Bad: The revolution in the wine world that occurred in the 1970s due to brilliant Californian wines fueled two major outcomes. First, and most important, it led to the demise of European superiority regarding wine quality. The wines of the Old World were still superb, but one could no longer look down on the New World's efforts to make wine. In some blind tastings, iconic European wines were coming second! This opened up the possibility that good wine could be made in many new locales, which was a marvelous development — good for all lovers of wine as wine production spread, sharing of knowledge flourished, and advanced techniques improved quality. Consumers were now blessed with many fine choices, even from their own regions.[64] The buy-local movement cheered!

The second major outcome of this Californian breakout was the common practice of putting the name of the grape(s) on the bottle's label along with the predominance of using just one grape in the wine. This latter practice is fine when that grape makes good wine . . . but not so when the grape is better used in blends. For example, in France, using the Cabernet Sauvignon by itself was *not* common. Yet in California, winemakers did it, and sometimes with great success. Some of these Cabs are very good, but many are lacking and somewhat one-dimensional. The same could be said for the Petit Sirahs, Sauvignon Blancs, Merlots,[65] and a number of others. The overall result was that much of the New World started following suit. The popularity of the varietal approach led to some bad decisions — and bad wine.

....

63   *Wine and food:* See the Appendices.

64   *Jug wines:* In early U.S. history, wines were named after French regions. Cheap, awful tasting wines were labeled Burgundy, Chablis, or Sauterne. Thankfully, this practice has largely been discontinued.

65   *Petrus:* One great wine made from Merlot, Château Petrus, is over 95% Merlot.

Many winemakers are trying to make varietal wines in soils and climates better suited to multiple grapes and blended wine. This is evidenced by the abundance of thin, neutral varietal wines on wine-store shelves. Sometimes people try to fix these poor wines by doctoring them. They start using shortcuts like adding oak chips (or oil extracts of oak), extra juice, artificial flavours, etc. They would have been better off simply blending different grapes to get a more natural, full-bodied, and better-tasting wine.

The other amusing development is the relatively recent phenomenon of French wine companies' (Italian too) starting to put the name of the grape on labels and following the varietal approach. Some companies located in Bordeaux, the finest region for blended wines in the known universe, are even marketing Cabernet Sauvignon! This is a desperate measure to regain market share lost since the rise of Australian, Californian, Chilean, and Argentinean varietal wines. I think the whole idea could backfire on the European companies; they should stick with what they do best. Wine drinkers are slowly becoming more sophisticated, and long-term, the key will always be quality. Winemakers should rely on the region's climate and soil type to determine whether they should make a varietal or a cuvée wine.

The Ugly: What could be uglier for a commercial enterprise than getting good results only one out of three years? Welcome to Burgundy's red wine producers. The Pinot Noir requires prolonged warm summers and a cool fall to ripen properly. When this happens, you will find great wines produced from that vintage. Unfortunately, Burgundy is inland and is often subject to cooler weather (Bordeaux, with its maritime influence, has much more stable climatic conditions). The result is that many reds from Burgundy bear little resemblance to the legendary wines that established the region's elevated reputation. This predicament is potentially repeated wherever winemakers follow the varietal approach. The winery is at the mercy of the weather (or diseases) to a greater extent than wineries with vineyards planted with different varieties. This is a nerve-racking consideration for every winery in the one-grape game. A lot of the time, quality is sacrificed in the varietal wines due to factors beyond the winemaker's control.

The other unfortunate outcome from the spectacular quality of the best varietals, the Riesling and Pinot Noir in particular, is that too many people have tried to emulate them elsewhere, sometimes in similar climates but in varying types of terroir. Problem is, both grapes are dramatically influenced by the soil composition, and each one is rather difficult to grow. The Riesling is a late ripener, yet it produces its best wine in cool climates — now there's a tough combination of growing factors! Meanwhile, the Pinot Noir is susceptible to rot due to the very tight formation of its grape bunches and its thin skins. Sometimes the harvest is spoiled before the grapes have a chance to fully ripen. And like Riesling, it too thrives in cooler climates but needs hot sun at the right time.

The wine world is littered with sad stories of people planting Riesling and Pinot Noir vines in unsuitable sites. Too much heat, and you get a boring wine. Too much cold, and you don't get a harvest. Too wet, and you get rotting grapes. Wrong soil, and you don't get any varietal, or typical, flavours. As mentioned, the Pinot Noir in particular has been called the heartbreak grape. You might as well say the same of the Riesling. A lot of time and a lot of money have been spent on chasing the elusive tastes of these two grapes.

Note: Luckily, there are inroads being made with both grapes. You'll find some fine Rieslings from Australia, New Zealand, Canada, and even New York State (not to mention the U.S. Northwest coast). The Pinot Noir has made a breakthrough in Oregon and cooler parts of California, where the true varietal character is showing through. New Zealand, Chile, and Canada are starting to get some pretty good results with this heartbreak grape as well.

III. QUANTITY VERSUS QUALITY

"What is the definition of a good wine? It should start and end with a smile." —William Sokolin, New York wine merchant

After deciding whether to produce a varietal or a blend, the second critical decision a winery faces is whether to go for quantity, making value-priced wines, or to make fine wines targeted at the discriminating wine lover. This decision is often linked to the size of the harvest. Either the grape grower/winemaker maximizes the amount of fruit per vine and acre, or he or she limits the amount of grapes to concentrate flavours (as a result of a smaller harvest). The factors influencing these different approaches include vineyard location, grape type, vine management, production techniques, and cost, among others.

(i) Value Wine

This type of wine is aimed at the majority of wine drinkers. The price will be in the $6 to $12 range, and surprisingly, the wine will often be very acceptable. I'd describe value wines as similar to the veneer floors one often sees in newer homes today. Superficially, they look the same as solid wood, but lacking the depth of solid-wood floors, they will wear out sooner. You'll notice scratches and chips in the floor. Still, for the money, veneer floors are often a great bargain and terrific value. So it is with good value wines.

VINEYARD CHOICES

The number-one decision, of course, is determining the site of the vineyard. The winery that wants large harvests will normally choose a warmer climate. In particular, they will look at *degree days*, which is a measurement of good growing areas; it is calculated by taking the average daily temperature and subtracting 50°F. All daily degree numbers are then totaled for May to October; 2,500 to 3,500 is considered the best range for grape production. Next, the winery considers terroir in its broadest sense — not only soil fertility but also slope grade, drainage, elevation, wind, rainfall, and the direction the vineyard faces (south-facing vineyards maximize the sun exposure in the northern hemisphere). All of these characteristics will affect the growing conditions for grapes.[66] The value-focused winery takes advantages of each aspect to increase the yield.

After choosing the vineyard site, the next essential choice is the type of grape to grow. The goal is to get a variety that yields abundant harvests while producing reasonable quality. The Chardonnay is a good example of a grape that is easily grown and ripens almost anywhere it is planted. It also produces palatable wine under most circumstances and is mostly in demand among the wine-buying public. A great value wine grape!

The quantity- (and quality-) focused winemaker must pay attention to the needs of the vine through the following phases, leading up to and including the harvest.

SIX ANNUAL STAGES OF GRAPE VINES

1. *Pruning:* The vine is cut back each winter, leaving only the desired amount of canes intact for the next spring. If you cut the plant back too much, you'll limit the growth and abundance of fruit. A vine left untended will end up putting too much energy into shoots, and there won't be enough nutrients left to ripen all the grape bunches.

2. *Bud break:* The canes now have a number of little nodules on them. Each nodule will suddenly form a small green shoot and start growing upward. The plant is now very vulnerable to frosts. A cold snap in the weather at this point could ruin the harvest.

....

66 *Other considerations:* The astute businessperson will also check out any potential *crop hazards* such as animals, including deer and bird life, as well as susceptibility to drought and disease in the area.

3. *Flowering:* Small flowers appear on the vines as the weather warms up. Once the vine is in full flower, the danger of frost damage is past.

4. *Fruit set:* The flowers are self-pollinating, and once pollination is complete, tiny clusters of seed-sized balls start to form. These are the beginnings of the grape bunches, which now begin to grow. Adequate water is critical from bud break through to the fruit set.

5. *Veraison:* In mid-season, the grapes change colour, a sign that they're truly starting to ripen. From this time to the harvest, it's important to manage the ratio of leaves to fruit; this is known as *canopy manage-ment.* Leaves are necessary for photosynthesis to occur (so sugar is produced); however, too many leaves will shade the grape bunches and adversely affect their health. Therefore, careful plant manage-ment is required to get vigorous fruit ripening.

6. *Harvest:* When the grapes reach maturity, they are harvested (the grape grower will measure the sugars and acids to determine this exact moment). Traditionally, grapes were picked by hand — an arduous and time-consuming process that is still practiced in top-quality wineries. In the quantity approach to grape growing, the grapes are harvested mechanically by large machines. This method is much faster and saves labour, and therefore money, for the producer. It's all part of the value equation (see Figs. 3 and 4).

Once the grapes are brought into the winery, the winemaking process begins. In value wine production, virtually all the grapes are included in the batch, whereas in fine wine operations, individual bunches are care-fully inspected and selected. The winemaker will add sugar if necessary to ensure alcohol levels are sufficient.[67] The wine is quickly fermented (usually in stainless steel tanks), aged briefly, and bottled as soon as possible for market. Note that white wines require the least aging. Reds often have longer aging, frequently in wood, before being released. Overall, the goal is to minimize the time that wine spends in the winery.

....

67 *Chaptalization:* This is the practice of adding sugar to the grape juice to make sure the alcohol reaches desired levels. Chaptalizing is practiced in both Germany and France. Used judiciously, it adds body to an otherwise thin wine.

GEOGRAPHIC DESIGNATION

Value wines usually have very little information on the label. You often see the name of the grape and/or some romantic or frivolous name. Information on the back label is limited as well (if there is one at all). The important piece of information is most often the geographical description. In general, value wines will be described as coming from a large region. For example, the label will say "Product of Spain," meaning the grapes could have come from anywhere in the country. A better wine will be more specific. A good value wine from America might say "California" on the label; an even better one would say "Sonoma County." This is a common theme around the world: The more *specific* the geographic description, the better the wine . . . in most cases.

A smaller geographic designation does not *guarantee* better quality; however, nine times out of ten, it's a good sign. The reason is that most countries will not allow you to put a clearly defined smaller region (e.g., a county name or town name) unless you adhere to strict vineyard and winemaking practices. These methods are all aimed at producing quality wines. Some of the regulations will specify whether you can irrigate (*not* permitted in some areas in France), the amount of grapes you can harvest per acre, fertilization practices, and even grape varieties permitted (the overriding logic being "the fewer grapes per acre, the better the wine produced"). It's interesting to note that in the Central Valley in California, where there's no limit on grape yields, a vineyard will sometimes produce over 20 tons of grapes. In fine wine regions, regulations often limit the production to between 2 and 4 tons! Quite a difference!

Specificity: In conclusion, the geographical designation *most* often reflects the strictness of the region's wine regulations. The smaller the defined area on the label, the tougher the rules and, hopefully, the better the wine inside the bottle. When you just see the phrase "Product of Country *X*" you are definitely looking at a value wine. I'll refer to this phenomenon as the *rule of specificity* in the last part of the book, "Label Logic."

CASH FLOW CONSIDERATIONS

It may already be obvious why many wineries specialize in value wines: They are usually more profitable. The winery gets more fruit per acre, uses less labour, streamlines production, and gets the product to market quickly. Compare this to a wine that is made from a harvest one quarter the size, handpicked, with carefully selected grapes (often difficult to grow), aged for a few years in expensive oak barrels, and finally bottled for sale. The implications are obvious in at least one sense: cash flow. The value wine companies have often sold their wine within a year of the harvest date. In any business, turning over your inventory as soon as possible is usually an advantage, and the wine industry is no exception.

Note that many of the largest wine companies engage in both the quantity and the quality approach. The best example is E. & J. Gallo, the largest wine company in the world. In the 1930s, two brothers, Ernest and Julio, started their winery in Central Valley, California. For years, they made "jug wines," which were bottled in large glass containers with a glass ring attached to the neck for holding onto; the quality was so poor you could not even call them *value wines.* Today, the Gallo company dominates the value wine category, yet it also makes some very fine wine. Using the profits from their years of appealing to the masses has allowed them to enter the world of quality wines — a nice evolution and a good story.

(ii) Fine Wine

I don't mean to suggest that wines can be clearly divided between value wines and fine wines. Quality is not a black and white thing; there's a lot of grey out there. It's better to think of wine quality as a continuum that goes all the way from poor quality, cheap plunk to wines that will enchant you with their beauty, fragrance, and flavours. As Robert Louis Stevenson once wrote, "Wine is bottled poetry" . . . and sometimes it can be! This part of the book will focus on those viticulturists and winemakers who are aiming at the quality end of the wine spectrum.

VINEYARD STRATEGIES

Makers of fine wine are striving for something quite different from their value wine counterparts. First, they're usually obsessed with grape health and carefully manage the six annual stages of grape vines. Their goal is to maximize the complex flavour components of the fruits. Not only do these winemakers want to balance sugars and acids, but they're also focused on *phenolics,* which are the compounds concentrated in the skins of the grape, including tannins. The grape growers' plans and efforts are focused on getting all these major elements in the right proportion at harvest time.

Not surprisingly, the fine wine route to quality is a rocky road. The best vineyards are usually on the worst land, as measured by the normal benchmarks of soil fertility. Almost without exception, the best wines come from grapes grown on poor soils and barren land. From the slate-laden soils of the Mosel to the gravel soils of Bordeaux to some of the thin, rocky soils of Sonoma, grapes flourish. These unfertile soils cause the vine to struggle a little and then send their roots downward to find water and important nutrients. The result is a more concentrated and complex fruit.[68]

Besides soil type, another important factor is *drainage.* Grapes grown in moist soils often produce too much foliage and yield watery, flavourless grapes. In fact, if there's too much moisture, the vine will become waterlogged and may even start to rot. Some experts believe that the structure of the soil is more important than its content. The relative importance of the soil's component parts versus its ability to drain well seems to be debatable, but one fact is not: Most great vineyards around the world are located *predominantly* on slopes. Some are fairly steep, like parts of Napa Valley, while others are like the gentle slopes of the Côte-d'Or region in Burgundy. This important truth seems to support the thesis that drainage is the most important factor for fine-wine vineyards. One way or the other, great grapes come from a variety of soils, mostly poor, that have excellent drainage capacity and where the vines are forced to struggle a little to survive.

....

68   *Deep roots:* The other advantage of deep roots is that the vine becomes more resilient and resistant to drought.

So how does the viticulturist assist in this struggle? Sometimes by just following the regulations of the region. For example, in many areas in France, irrigation is forbidden and any fertilization is limited to the beginning of the growing season. These rules do exactly what the poor soils do in the best vineyards: They limit the fruit by allowing the vine to struggle and allow nature to run its course. The overall goal is to produce a smaller harvest with more flavourful and concentrated grapes. Although other areas of the world do not have the same strict regulations as those found in the French wine laws, winemakers will often mimic traditional practices to encourage the same results as those of the iconic wines produced in regions like Burgundy and Bordeaux.

Thinning: A common practice employed to achieve greater quality is to *thin out* the vines. Early in the season, as the canes of the vine begin to send shoots out, the viticulturist/farmer will limit their number.[69] The farmer pulls off the smallest ones, thereby leaving only the primary shoot to mature and bear fruit. Sometimes individual shoots are too close together, so some are removed to allow better spacing; this prevents overcrowding of the subsequent grape clusters.

The second type of thinning is accomplished by actually chopping off grape bunches in mid-season. The farmer determines the healthiest looking bunches on a shoot (usually two to three grape bunches per shoot) and cuts off the others so the healthier ones can thrive and fully mature. This type of thinning is carried out just before *veraison,* when the grapes are starting to truly ripen. As the old saying among winemakers states, "It's not the grapes you harvest that makes a great wine; it's the ones you left on the vineyard floor." The whole goal is to find the balance between grape quantity and grape quality. Too many grapes, and you end up with thin-tasting, inferior fruit (sometimes the grapes will not even ripen fully). Too little of a harvest may allow too much sugar to be concentrated in each grape; this could lead to overly alcoholic wines, which is often not desirable, not to mention the fact that the winery also has less wine to make *and sell* (remember that a winery is also a business!). Thinning requires a delicate balance for the

....

69   *Thinning:* This is not the same as pruning, which is done in the dead of winter.

winemaker . . . the yin and yang, again (i.e., quantity versus quality or business versus focus on craft).

WINEMAKING

Once the farmer hands over top quality grapes to the winemaker, a number of critical decisions must be made in the winery to enhance quality.

Fermentation: The grapes, after careful sorting, are quickly put into fermentation tanks to allow the yeast (either natural or cultivated yeast, depending on the winery's philosophy) to convert the sugars into alcohol. Sometimes the winemaker will ferment in large wooden barrels to get a stronger oak presence in the wine; however, the majority of wineries now use stainless steel tanks for fermentation. For reds (which are fermented on the skins to extract colour and phenolics), the winemaker must decide how long to leave the skin in contact with the grape contents before the grapes are pressed. After the wine juice is fermented, the new wine needs to be readied for the market.

Aging: The next important stage involves aging the wine before bottling. For whites, this usually requires a transfer into another stainless steel vat, where the wine rests. Certain white wines are aged on top of the *lees,* the dead yeast cells left after fermentation. This will add a little depth, which is sometimes desirable. Very few white wines, with the exception of Chardonnay-based whites, will be aged in oak barrels. Meanwhile, many newly fermented reds are transferred to oak barrels for a few months or more (some great, concentrated reds will be left in oak for more than three years before bottling). The oak aging allows the wine to breathe a little, resulting in slight oxidization, which softens the tannins from grape skins and opens up the other flavour components. The wine will also pick up some different tannins from the oak. If everything is timed properly, the mature red wine will have gained a more rounded flavour, a silkier texture, and a hint of vanilla and oak aroma, thereby making a seductive brew!

VINEYARD SPECIFICITY

This final part of the "Fine Wines" section is not directly related to winemaking techniques; it involves a decision regarding which vineyard source to use for the final wine. The winemaker can use wines made from grapes grown in many vineyards or narrow down the selection to certain smaller parcels of land. When a specific vineyard name appears on a label, it usually denotes a higher-quality product. The wine is likely from a vineyard with special quality attributes in its grape harvest. In France, and increasingly in many other countries, the vineyard name on the label is allowed only under very strict circumstances or regulations. Most wineries employ this labeling practice only with their most prized vineyards. More often than not, the wine will be a company flagship: their most prized or ambitious wine.

IV. LABEL LOGIC

"Every glass of wine we drink represents a whole year of vineyard cultivation and perhaps several years of effort in the winery. . . . Yet most of us throw it away, straight down our throats, without even trying to read it." —Jancis Robinson, Jancis Robinson Wine Course (PBS)

The final part of the book deals with all the rules and regulations affecting information permitted on wine labels. The laws are complex and can be confusing, but a certain common logic governs the information on most countries' wine labels. I will try to identify important words and descriptions that will give you insight into the wine inside the bottle. Much of this information is similar to the clues a detective must uncover in order to solve a mystery. For me, there's no better mystery to investigate!

Why are there wine "laws" in the first place? We'll look at the history and some of the reasons for the establishment of the earliest regulations in France. It's important to note that the label rules in France and the rest of Old Europe (essentially Italy, Spain, Portugal, Greece, and, importantly, Germany) have set a world standard. It is in these European Union countries that common label logic has emerged. We'll also see how Germany has taken a slightly more scientific route in its label regulations. Finally, we'll look at North America and a few other New World countries to see what parts of the European system of rules have been adopted (so far).

I'm willing to predict that one day, all wine regions will use labeling laws that are essentially a synthesis of the French and German rules. The French system is focused mainly on the type of grape(s), its origin, and winemaking practices. The label is a combination of a birth certificate and a certification of pedigree, with a few details on the wine's upbringing. Meanwhile, the Germans have centered more on the grape's ripeness at harvest and overall quality control of the wine itself. Merge the two systems, and we'd have labeling rules that I believe would be best for today's consumer — a consumer who increasingly demands transparency, traceability, integrity, and, of course, quality.

(i) The Origin of Label Laws in France

Earlier in the book, we discussed the beginnings of wine regulations in 14th-century France, when the Duke of Burgundy banished the Gamay grape. His decree was based on the belief that the Pinot Noir made better wine. Farmers preferred the Gamay because it was easier to grow and yielded more abundant harvests. They were obviously motivated by profit. The Duke would have none of that; he wanted better wine. To some extent, this same battle of quantity versus quality is at the core of many wine regulations.

The most famous attempt to classify wines occurred in 1855 in Bordeaux. This classification was made for that year's Universal Exhibition in Paris and was based mainly upon the prices certain wines were expected to get (or were already getting in Bordeaux) at the exhibition. The wines were divided into five categories, or "growths." The best were known as the *first growths*[70] (*grands crus* or *premiers crus* in French), and the last category was the fifth growths (*cinquièmes crus*). This was the start of a system to rate wines for the market. More-stringent classifications and regulations were soon to follow for some very good reasons.

Two developments in the mid-19th century caused major changes in the French wine industry. First was the devastation of the vineyards because of the invasion of that destructive North American louse. The phylloxera out-break ruined many wineries and many people's livelihoods. Many wineries were forced to resort to some radical methods to maintain production.

....

70   *Grands crus:* In 1855, there were only four "first growths." They are among the greatest reds in the world: Châteaux Haut-Brion, Lafite-Rothschild, Latour, and Margaux. In 1973, Château Mouton Rothschild was added to the *grands crus* — the *only* change to the 1855 classification!

The second development was significant scientific progress that helped some of these beleaguered businesses. Through science, winemakers understood the workings of fermentation. Winemakers also learned about various "additives" that could improve quality and/or the quantity of wine from the grape harvest. However, not all wineries used the new scientific knowledge honourably. Some began to employ fraudulent techniques to stay in business. Soon they started adding more sugar, water, colouring (even blood), industrial alcohol, inferior hybrid grapes, imported juice from Algeria . . . anything to increase production. There's an old wine joke about a dying vintner calling his son to his bed, where he whispers, "Son, you can make the wine from grapes too." The scale of this fraud was quite extensive during this period in Europe; often, it was the only way to stay in business.

Of course, not all European winemakers resorted to unscrupulous practices. The better wineries used newfound discoveries to overcome product difficulties. Most replanted their vineyards by grafting the European *vinifera* vines onto the phylloxera-resistant North American roots and made "real" natural wine. The grafting approach would take a lot of time (3 to 5 years), as it required waiting for the vines to mature and to slowly begin producing good fruit. In the meantime, the artificial winemaking practices continued well into the 20th century, filling the demand of a thirsty public. Finally, the French government intervened to regulate the production, and other European countries soon followed. New laws were established in France to protect the honest winemakers *and* the consumer.

Eventually, the French created rules that guaranteed the origin and type of grape along with winemaking best practices in defined wine-producing regions. Through many years of debate, they managed to agree upon which grape was best-suited to various regions. This matching of grape to geography (actually, soil) is at the heart of a terroir-centric philosophy. The French fervently believe certain grapes do better in some soils and certain climates. They have enshrined this thinking into their wine laws.[71]

....

71 *Static laws:* Some people argue the European rules prevent creativity. Being restricted to a limited number of grapes might be a mistake. Experimentation with other varieties may result in a few surprises.

The ultimate outcome of this history in France is the *Appellation d'Origine Contrôlée (AOC)*. This literally translates as "the name of origin certified." The AOC is used only for the highest-quality wine regions. The consumer now knows that the grapes were grown in the named region and which grape variety was used.[72] As the regulations evolved, the AOC designation came to mean that there were controls for the permitted yield, vineyard irrigation, grape-vine growing techniques, and winemaking practices. An AOC label also ensures the wine conforms to the region's basic style; this last criterion is confirmed by chemical analysis and tastings by experts. Besides the AOC designation, the French also have other grades describing wines that are, for the most part, of lesser quality and require less rigorous production practices (see subsection V, "GEOGRAPHIC TABLES AND WINE LABEL SAMPLES").

(ii) Germany's System

It's almost a miracle that grapes can even grow in vineyards that are near latitudes of 50 degrees (that's equivalent to Newfoundland!). Yet it is here, in Germany, that the world's most delicate Rieslings are made. The extreme northerly location obviously means cold weather; however, it does have one big advantage: longer sunlight hours each summer day. The better vineyards are mostly south-facing which, along with some hard work, allows the late-ripening Riesling grape to fully ripen. The result is an intensely aromatic, fruity tasting wine with firmly balanced acidity. Such wines are often compared to heavenly nectar (by me!).

Germany's wine regulations and its labeling system are mainly concerned with sugar. And no wonder that the Germans are obsessed with sugar levels in their grapes! After all, achieving ripeness in this northerly region is incredibly difficult. As the grape matures into the fall, the sugar level increases dramatically. Inversely, the acids start to decline. The trick is to pick the grapes when the sugars and acids are at the right proportions.

....

72   *Grape name:* In the best regions in France, it is *forbidden* to put the grape name on the label (Alsace being a notable exception) for AOC wines.

Other than this focus on sugar, the Germans also analyze their wines carefully. Each batch of wine goes to a quality-control center, where it is analyzed chemically and tasted. The wine label even has an official AP number (Amtliche Prüfungsnummer), which indicates the place and the date where the analysis was conducted. The analysis is all very methodical and scientific.

In Germany, the Riesling is the most-prized grape variety. It's often regarded as a rival to Chardonnay as the finest white grape. However, two other white grape varieties are more widely planted: the Müller-Thurgau and the Sylvaner. Both make good wines and are easier to grow than the Riesling. Germany also makes some good reds. My favourite is made from the grape known in Germany as *Spätburgunder* (in Burgundy, called the *Pinot Noir* grape). This red wine is often quite light in weight and colour, but it has nice spicy, fruity, and subtle flavours.

The best of the German wines are the ones made from fully ripened grapes. When these wines are made, no sugar needs to be added (the process known as *chaptalizing*[73]). These wines are allowed to be labeled *"Qualitätswein mit Prädikat"* (QmP), which, roughly translated, means "wine with special attributes." There are three other categories of lesser quality, known as *Tafelwein, Landwein* and *Qualitätswein bestimmter Anbaugebiete* (QbA). The first two categories are the lowest in quality; such wines have neutral character and are made from grapes that are allowed to come from anywhere in the country. The QbA wines are from one of the official regions in Germany, of which there are thirteen. The two most important regions are the Mosel River area (including its two tributaries, the Ruwer and Saar rivers) and the upper Rhine River area known as the *Rheingau.* The very best of the German white wines are made from these two great river regions.

Let's look at how the top wines are defined on the label. At first glance, German wine labels appear fairly complicated. However, they follow a strict logic. I will focus on the QmP (special quality) wines and how they're

---

73   *Chaptalization:* The adding of sugar to the juice before fermentation is actually forbidden in the top-quality wines in most countries.

labeled. The system is all based on the *different levels of grape ripeness* used for each subcategory.

Here are the key German terms used and a brief description of the grape's condition and the wines produced from them:

1. *Kabinett:* This is the lowest degree of ripeness (i.e., lowest sugar at harvest). The term, which translates roughly as "cabinet," originally referred to a wine that was stored in someone's special storage area for guests. These wines are mostly dry, light, and refreshing.

2. *Spätlese:* The next level of ripeness comes from more mature grapes. The term means "late harvest." Leaving the grapes on the vine later into the autumn allows for greater sugar content. These wines have a little residual sugar and are fuller tasting than Kabinett wines.

3. *Auslese:* This term refers to "selected" grapes. This means the grape bunches were carefully chosen for their superior ripeness. This is a complex, sweet wine that will improve with some aging.

4. *Beerenauslese:* Now we're looking at meticulously selected *individual grapes. Beeren* means "berries," so each grape within a bunch is singled out for this wine. Sound labour intensive? It is, and it's expensive. The wines are sweet and age very well. Note: Many of the grapes are actually botrytized (affected by noble rot), which helps to concentrate sugar.

5. *Trockenbeerenauslese: Trocken* is a term that you'll see on other German wine labels to describe *dry* wines. In this case, it refers to "carefully selected, fully dried grapes." In other words, almost raisins! This is a very sweet wine similar to French Sauternes, but it's more complex in my view. Amazing as dessert! This wine is also very rare and will age for years.

The preceding terms, a mouthful in themselves, are actually quite easy to learn. The English equivalents appear on many New World labels. In some countries, those terms are actually part of their wine-labeling regulations; i.e., if a label says, "late harvest," then legally, the grapes must have reached a certain level of sugars by harvest time. However, the English versions of the terms are most often used as descriptors and are not regulated.

EISWEIN (ICE WINE)

This is a category of sweet wines that is dear to my heart. Coming from Canada, the largest producer of this elixir, it is a wine with which I've had considerable experience. *Eiswein,* as it's called in Germany, is made from grapes that have frozen on the vine. They are then pressed, also under frozen conditions. The juice comes out extremely concentrated in sugars and acids, while much of the water remains behind as ice crystals.

Years ago, I made ice wine with one of my business associates, and the experience was a little comical. After putting the frozen fruit into an old wooden press, we firmly turned the screw mechanism. Nothing happened! After about twenty minutes, the grape juice finally started to trickle out of the press; it was like thick corn syrup. It took three days to press just four bushels. Luckily, the wine slowly fermented (too much sugar *inhibits* yeast activity, causing fermentation to stall at times). We managed to keep the process going, and the wine turned out to be delicious. Like a German ice wine, the aromas were immensely intense with hints of flowers, oranges, honey, and apricots. The flavours lingered on for what seemed like hours — probably a result of our hard work!

FINAL THOUGHTS ON GERMAN WINES AND LABELS

The white wines of Germany are largely ignored by the North American consumer. That's a shame, because these wines are among the most enjoyable wine experiences available . . . and they represent *great* value. The misconception that many people have is that all German wines are too sweet. Well, sweet they are but not like candy. German white wines have a tremendous acidic backbone that mitigates the sweetness. They burst onto your tongue with delicate, intense fruit flavours and magically finish in balance. In fact, some even taste virtually dry as they linger in your throat. The combination of the citrus and flowery aromas, fruit flavours, and refreshingly clean aftertaste is marvelous.

As I noted before, another terrific aspect of German wines is their alcoholic levels. In general, they're very low. This is a particularly pleasant contrast to today's fruit-bomb, alcohol-laden wines that are completely lacking in subtlety yet are so popular in many markets. One of my favourite wine experiences was sitting in Bernkastel on a hot day along the Mosel River, with my family, sipping on a wine from the Doctor vineyard. It was only *7% alcohol*, which is similar to some beers! An amazing symphony of bouquets and tastes that left you almost breathless in admiration of its true finesse. A second bottle was just as pleasing!

Finally, and most importantly for our understanding of label logic, the German wine labels offer the most detail in comparison to any other country. They include not only the quality designation we've discussed for Qualitätswein mit Prädikat wines but also the name of the region, the village, and the vineyard. Plus there's the producer's name and address, along with information about where the wine was bottled. And to top it all off, there's the A.P. number, which helps authorities trace when and where the wine was tested. Many aspects of the German labeling regulations serve as a good model for all emerging wine regions and countries.

(iii) Lessons Learned in the New World[74]

The main feature of European wine label regulations adopted throughout the New World is identification of the grape-growing region. This geographic *appellation of origin* is a central element of the majority of labels on quality wines. For wine lovers, it is the main clue for identifying where certain grapes were grown and consequently allows us to judge wines from one area to a greater degree. For instance, once a winery makes an excellent wine from the Pinot Noir grape grown in a precise region, like the Willamette Valley in Oregon, then you can guess that other producers in that area might also match that result. Armed with this assumption, you can look for other Pinot Noirs from that same valley. It is the origin of the grapes that is most important. Of course, seasonal fluctuations, vine-growing skills, and winemaking technique all come into play, but the terroir is the most important factor. This is why all the better New World regions have adopted this one key practice in labeling their wines: appellation of origin.

....

74 *New World:* For the purposes of this book I will only discuss United States, Canada, Australia, Chile, New Zealand and South Africa.

BASIC INFORMATION

Other than the most basic information, like alcohol content and bottle size, most countries have incorporated rules governing grape names on the label, appellations, the vintage date, and the producer's name and address.

Let's look at what each of these four pieces of information might require to be approved for use (by the governing bodies):

1. *Grape name:* All countries require the winery to use a minimum amount of a grape in the wine in order for it to use the name of *that* grape on the label. For example, if the wine is a varietal called "Chardonnay," then at least 75% of the grapes used must be Chardonnay. The percentages change, country by country, but a *minimum* is normally defined. If more than one grape is included, then they normally have to be listed in order of proportion: The grape used the most in the blend is listed first, and so on.

2. *Appellation:* Sometimes an area will allow just the use of a certain place name or term to describe the wine, versus denoting grape varieties. Thus, you'll often see a geographical area like "Sonoma" on a label. In other cases, the label will contain a defined term. For example, in California, producers can use the term *Meritage* if the wine is made from a blend of traditional Bordeaux grapes (whites from Sauvignon Blanc and Sémillon; reds mainly from Cabernet Sauvignon, Merlot, and Cabernet Franc and sometimes Petit Verdot and Malbec). The use of a geographical area is controlled in most countries now.

3. *Vintage date:* The date when the grapes were grown is often a valuable piece of information, especially in regions of the world where seasons fluctuate considerably. For example, if you know a wine was made from grapes grown during a cool and rainy summer, then the wine *may* be thinner, more acidic, and more one-dimensional than normal. Some regions like the Napa Valley are blessed with mostly uniformly warm summers, so the vintage date is less critical for judging the wine inside the bottle.

4. *Producer's name:* The maker of the wine is a very important piece of information in any country, Old or New World. There are certain

names you can trust. In Italy, for example, Antinori will almost always make good wines. In the New World, the producer's name is often the best information you'll get on the label, because there are fewer regulations in these countries. This greater flexibility for wineries often leads to wines of varying quality being made in the very same region. For example, a winemaker in Australia may blend poorer grape varieties into his or her Shiraz or neglect to thin out the vines; without thinning, too many grapes grow, and they end up lacking true varietal character. In many instances, you learn, by trial and error, which producers follow best practices. Otherwise, you can check out wine competition results; read reviews on the Internet, in newspapers, and in magazines; or go to your local wine store to get expert advice (assuming a knowledgeable consultant is on staff).

NOTABLE TERMS AND CLUES

As I mentioned, many of the New World countries have incorporated some of the descriptive terms used in Europe. Some of these terms are governed by the country's wine bodies, while others are simply used by certain producers as a general indication of their wine. Again, with the reputable producers, these terms are accurate descriptors of the style and/or quality of the wine inside. Here are some examples of meaningful terms and clues you'll find on wine labels:

1. *Late Harvest:* This term mimics the German QmP hierarchy of quality wines. It means the grapes were picked later and therefore are (or should be) higher in sugar. The wine, as a result, will be sweeter — suitable for a dessert accompaniment or a match for some strong cheeses. In Canada, this term is regulated so that you can be assured that the grapes must have reached a certain sugar level prior to picking.

2. *Reserve:* The better winemakers use this term only for their best wines. In Europe, it often means that the grapes achieved greater ripeness, that the wines were not chaptalized, and/or that the wines were aged longer in oak barrels prior to bottling.

3. *Traditional Method (Méthode Champenoise):* This term describes sparkling wines that are made in the same way as real Champagne — i.e., following the same techniques used to make sparkling wines *in* Champagne. Simply put, it means that the wine was made and then re-fermented in the bottle, with the CO_2 trapped inside, and that the bottle you are buying is that same bottle (see "Champagne" in Section One).

4. *Proprietor's Selection:* This term usually means the wine was made from "selected" grapes (just like top German wines) or from the winery owner's favourite sites or regions. If the producer is honest, then this should generally be a good wine — a wine the owner is proud to sell.

5. *Botrytized:* These wines have been attacked by the fungus that dries out the grapes and concentrate sugars. Look for them to be delicious dessert wines and much sweeter than "late harvest."

6. *Vineyard Name:* When the label includes the name of the actual vineyard from which the grapes came, you can often assume it's a quality wine (again, from the best producers). This clue brings us back to my general rule of specificity: The smaller the appellation of origin, the better the wine. This is not an absolute axiom, but it is, for the most part, a good indicator of quality. A good analogy to my rule of specificity is found in some foods like meats and cheeses. A mediocre cheese might simply say "Product of U.S.A.," while a top cheese will say "Product of Wisconsin" or might even give a town or county name. In Europe, you are not allowed to put the vineyard name on the label, in most cases, *unless* the grapes were grown and wine was made in a carefully prescribed way. As for the New World, the use of a vineyard name is usually uncontrolled, but it's still an important clue as to whether the wine is top quality (if it was made by good producers).

7. *Estate Bottled:* This is another term that can be a very important indicator of top-quality wines. It usually means that the wines were made from grapes grown, fermented, aged, and bottled at the same property. This shows that the producer looked after every stage of the winemaking process. If producer cared that much, then you

can usually bet the end result is superior to other wines from the same area.

8. *Medals:* A lot of companies enjoy entering their wines into competitions (these contests are often part of large wine festivals). Medals are awarded based on judging by expert panels (often, a second set of awards called "people's choice" is based on the favourites of the general public at the festival). Many wineries put a copy of the medals they've win onto the label. A top medal can be a good indication of the wine's quality, especially from the better-known competitions.

A FINAL WORD ON REGULATIONS

Over the centuries, wine-related rules have evolved to prevent fraud and improve quality. Most countries have been successful in the first case and have been able to stop unscrupulous practices. Regarding the second goal of improving quality, the results have been mixed and open to question in some instances.

A good example of how regulations can actually *prevent* quality improvement is the story of wines called the *Super Tuscans* in Italy. A number of producers started using grape varieties not approved in Tuscany, but their wine exceeded the quality of the wines made in the traditional and regulated manner. The label rules prevented them from using the appellation of origin on their labels. For example, even though the wines were made in the famous Chianti region, the producers could only put "Product of Italy" on the label. This restriction is a good example of how wine laws can hold back progress (but to their credit, Italian wine bodies are now open to some changes, and their wine regulations are evolving).

Even in France, certain wine regulations could be hindering the creation of more-consistent quality wine. For example, in some of the top wine regions, irrigation is forbidden — "let the vine struggle to survive, and it will push its roots deeper and produce rich, concentrated berries," goes the thinking. But in a drought? Nonsense! Another example is the restriction on fertilizers. Surely a judicious, scientific application of fertilizers would ensure consistent harvests. Sometimes these regulations tend to ossify rather than nurture innovation and advancements in quality.

A country that's turning heads in the wine world is Australia. The country may have Australian Rules for Rugby, but hardly any rules face the wine industry. Wine labels in Australia have very few *mandatory* requirements. Grape types and blends are left to the winemaker's imagination. As a result, winemakers produce some interesting blends, like Cabernet Sauvignon with Shiraz, and Chardonnay with Sauvignon Blanc. Sometimes these non-traditional blends work and the wines are delicious!

The best that can be said for wine regulations is that they can give you a lot of confidence regarding the wine inside the bottle. For me, they're like a guarantee not of quality but of authenticity, transparency, and, most often, the winemaker's aspirations. The rules are *not* a rating or judgment of the wine but rather a definition of origin, grape varieties, viticulture practices, and winemaking techniques. Pretty important stuff! The wine itself will, for the most part, be similar to other wines made in the same area because the wines are made according to the same procedures. That's pretty comforting information when you're buying a wine based only on reading the label (and checking out the price!). Appropriate regulations are essentially an honest attempt at quality control. That's a pretty good intention — and good for us, the consumers.

V. GEOGRAPHIC TABLES AND WINE LABEL SAMPLES

"The point of drinking wine is . . . to taste sunlight trapped in a bottle and remember some stony slope in Tuscany or a village by the Gironde." — *Sir John Mortimer, English barrister and writer*

To clarify the information and regulation surrounding label content, this subsection will outline the basic regulations in twelve countries. The first six represent the main countries in Europe. I've set out the basic designations in a table, which is followed by some examples of real labels with helpful interpretations. Just remember that the *rule of specificity* is the secret in the quality hierarchy in the Old World: the smaller the region defined on the label, the tougher the regulations regarding grape-growing and winemaking . . . and as the theory goes, a better wine is the result.

The next six countries I'll discuss are all in the so-called New World. I'll treat them individually, because the regulations are not nearly as uniform as in Europe (overall, there are fewer restrictions/regulations) and therefore don't work well in a table. However, there are many parallels to Europe, as you'll see . . . and in my opinion, the trend will be to adopt more and more of the Old World approach to labeling and regulating wineries. Today's consumer will demand it.

OLD WORLD LABELS

The guiding breakdown of geographic areas on wine labels, from larger to smaller regions (shown below), follows a typical pattern for most European countries:

1. *Table Wine:* A wine that is defined simply as the product of a country (e.g., "Product of Spain"). The grapes can come from anywhere in that country.

2. *Regional or Land Wine:* These regions are usually quite large and are subject to tougher regulations than table wines.

3. *Quality Wine Produced in Specific Regions:* These wines come from much smaller areas, and the producers are more committed to making quality wine.

4. *Quality Wines with Special Attributes:* We're now looking at a very small percentage of wines. These are generally terrific — and very expensive — wines. Most of the work in the vineyard is by hand, and the attention to detail is very high in all aspects of the operation, including meticulous care of the wine from fermentation to bottling to aging (in many cases, the bottles are held in storage and are not even released to the public for a number of years).

This table shows the terms used in each country.

EUROPEAN WINE LABEL TABLE WITH FOREIGN LANGUAGES

Least inexpensive → Medium → Most expensive

Country	Table Wine	Land Wine	Quality Region	Special Attributes
France	*Vin de Table*	*Vin de Pays*	*Appellation d'Origine Contrôlée (AOC)*[75]	*Premier Cru* or *Grand Cru*
Germany	*Tafelwein*	*Landwein*	*Qualitätswein bestimmter Anbaugebiete (QbA)*	*Qualitätswein mit Prädikat (QmP)*
Italy	*Vino da Tavola*	*Indicazione Geografica Tipica (IGT)*	*Denominazione di Origine Controllata (DOG)*	*Denominazione di Origine Controllata e Garantita (DOCG)*
Portugal	*Vinho de Mesa*	*Vinho Regional*	*Indicação de Proveniência regulamentada (IPR)*	*Denominação de Origem Controlada (DOC)*
Spain	*Vino de Mesa*	*Vino de la Tierra*	*Denominación de Origen (DO)*	*Denominación de Origen Calificada (DOC)*
Greece	*Epitrapezios Oenos*	*Topikos Oenos*	*Onomasía Proeléfseos Anotéras Piótitos (OPAP)*	*Onomasía Proeléfseos Eleghoméni (OPE)*

75 VDQS: France has an AOC-in-training designation called Vin Délimité de Qualité Supérieure (VDQS). It generally indicates a good-quality region — some such wines become AOCs.

FRENCH LABELS

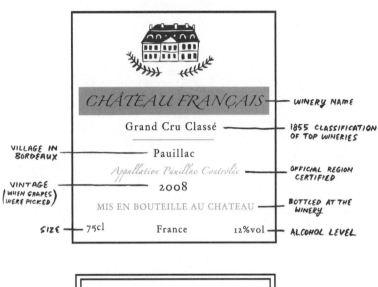

WINERY NAME

CHÂTEAU FRANÇAIS

1855 CLASSIFICATION OF TOP WINERIES

Grand Cru Classé

VILLAGE IN BORDEAUX

Pauillac

Appallation Pauillac Contrôlée — OFFICIAL REGION CERTIFIED

VINTAGE (WHEN GRAPES WERE PICKED)

2008

MIS EN BOUTEILLE AU CHATEAU — BOTTLED AT THE WINERY

SIZE — 75cl   France   12%vol — ALCOHOL LEVEL

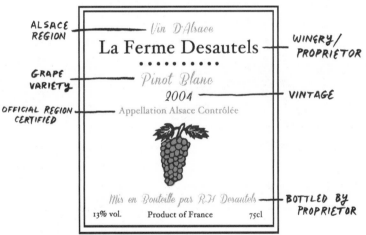

ALSACE REGION

Vin D'Alsace

La Ferme Desautels — WINERY/ PROPRIETOR

GRAPE VARIETY

Pinot Blanc

2004 — VINTAGE

OFFICIAL REGION CERTIFIED — Appellation Alsace Contrôlée

Mis en Bouteille par R.H Desautels — BOTTLED BY PROPRIETOR

13% vol.   Product of France   75cl

GERMAN LABEL

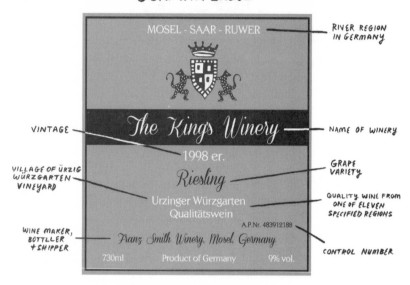

VINTAGE

VILLAGE OF ÜRZIG
WÜRZGARTEN
VINEYARD

WINE MAKER,
BOTTLLER
+ SHIPPER

RIVER REGION
IN GERMANY

NAME OF WINERY

GRAPE
VARIETY

QUALITY WINE FROM
ONE OF ELEVEN
SPECIFIED REGIONS

CONTROL NUMBER

MOSEL - SAAR - RUWER

The Kings Winery

1998 er.

Riesling

Urzinger Würzgarten
Qualitätswein

A.P.Nr. 483912188

Franz Smith Winery, Mosel, Germany

730ml   Product of Germany   9% vol.

ITALIAN LABEL

VINTAGE

INDICATION OF
AUTHENTICITY

GRAPE + AREA IN
NORTHERN ITALY

PRODUCER +
BOTTLER
IN ITALY

2010

Barbera d'Asti

Denominazione do Origine Controllata

IMBOTTIGLIATO DA
Roberto Desoto d'Asti
Calamandrana (Italia)

14 % vol.   75 cle

Geography and Grapes: As you may have noticed in the label examples, one must know the approved grapes in each geographical region. This makes most European wine labels difficult to read. Only in some countries, like Germany, does the name(s) of the grape(s) appear on the label.

Therefore, a bit of homework is required. Learning the approved grapes isn't as difficult as it first seems. Before long, you'll know Burgundy wines are almost all varietal wines made with one grape; the best whites are made with Chardonnay, and the best reds are made with Pinot Noir. Or you're recognize that although a Chianti red allows up to thirteen grape varieties, the Sangiovese is the predominant grape. A good resource book by someone like Robert Parker or Jancis Robinson will help you immensely in learning or looking up this information.

NEW WORLD LABELS

In North America and most other countries new to the world of wine (when compared to thousands of years in Europe), the overriding logic says to put the name of the grape(s) on the label. This practice makes it clear whether the wine is either a single grape (varietal wine) or a blend of grapes. The winemaker's quality aspirations are much harder to determine, as there are fewer regulations and less emphasis on defining geographical regions — although this is changing.

Let's look at six countries that produce the greatest amount of quality New World wine.

United States: The regulations in the U.S. are controlled by the Alcohol and Tobacco Tax and Trade Bureau, whose role is to prevent fraud and protect the public. Each winery, prior to selling its products, must submit its labels to the Bureau; if a label meets the standards, then it receives a Certificate of Label Approval. This gives consumers assurance that the wine is properly described on the bottle.

Some of the important basic rules concern grape names on the label. If the wine is a varietal, then it must have a minimum of 75% of the named grape in the wine (Oregon is 90%) and include an appellation of origin; the wine must also have 75% of its grapes from that named region as well (some states set a higher percentage). Significantly, there is *no* sensory tasting as part of the Bureau's controls.

In 1981, the U.S. started to establish American Viticultural Areas (AVAs). These subregions were defined by distinctive climatic conditions, topography, and/or soil composition. The goal was to define basic terroirs and attempt to better indicate the character of wine in a manner similar to the European system. The regulations for AVA wines are more stringent than for other wines in America. For example, 85% of the grapes in the wine must come from the AVA that is named on the label, and varietals must be made from a minimum of 85% of the named grape variety. Today, there are approximately 200 AVAs across the country.

Canada: The majority of quality wines come from two provinces, Ontario and British Columbia. The Vintners' Quality Alliance (VQA) is based upon a number of regulations similar to French appellation laws. VQA rules require that 85% of grapes come from the region named. There are also lesser "provincial" wines, which only require that the grapes come from anywhere within the whole province. Note: VQA wines are all subject to chemical analysis and expert tasting panels.

Canada has also adopted some regulations similar to the QmP wines of Germany. Minimum sugar levels in harvested grapes are required for using terms like "late harvest," "botrytis affected" and "ice wine" (the latter wine put Canada on the international wine map). There's no doubt that ice wine is extraordinary, but it has helped to distort the country's image for making quality table wines. Rieslings and Chardonnays from Ontario are very good (I also believe the Pinot Noir will eventually be a winner). Finally, British Columbia's Okanagan Valley produces many outstanding reds, particularly Merlot, along with some very good whites.

Australia: In the past, it was not uncommon for Australian wines to be blends of grapes from different production regions in the country. This blending is now done mainly for the value wines. The industry is start-ing to regulate some information on labels, yet very little is mandatory. Despite this lack of regulations, Australian wine labels tend to have more information on them than almost anywhere else in the world![76] Grape varieties, vineyard location, and winemaker names all make their way onto these labels. All Australian states are making very good wines, with South

----

76   Back Labels: Australians have a tendency to put a lot of information about their wines on the back label, including the winemaker's name and signature. Some of the stories on the labels are rather amusing.

Australia leading the way (Barossa Valley, McLaren Vale, and Coonawarra are the best subregions).

For years, the Aussie consumer relied mainly on grape names and the reputation of the producer to determine style and quality of their wines. This is changing, and an appellation system is developing along with a "Label Integrity Program." If a grape name appears, it must comprise 85% of the grapes in the wine, and the same percentage must be derived from the geographical area defined on the label. There's also a hierarchy of regions, from several states to over 60 subregions. These are tracking the distinct styles from many of Australia's regions. A Cabernet Sauvignon from the Coonawarra is very different from one from the Hunter Valley (as are the soil and the climate). I'm willing to bet that one day, the Aussie winemaker will be subject to more mandatory rules and regulations. Just wait, mate!

New Zealand: In the early days, producers in New Zealand used German and French regions to name their wines. Much like the U.S., they made "Chablis," "Champagne," "Burgundy," etc. In 1994, New Zealand started to define borders for good wine growing regions. The regulations require 100% of the grapes to be grown and made in the region named on a label. Over 75% of the country's wines are white, most famously from the Sauvignon Blanc (Note: New Zealand is the world leader in the use of screw caps; fewer than 25% of their wines now use corks!).

Chile: The *Denominación de Origen* (DO) was established in 2002. Chile has five main regions that, in turn, are broken down into smaller subregions. The wine label rules and regulations are still emerging, but expect them to reflect European laws. Some of the best available wine values in North America come from Chile. In particular, look for reds from the Maipo and Colchagua valleys.

South Africa: This country was one of the earliest in the New World to establish an appellation system. In 1973, the Wine of Origin (WO) classification system created a hierarchy of areas, from larger to smaller, as follows: *region, district,* and *ward.* The majority of South African exports are made in the Coastal Region from the districts of Paarl and Stellenbosch. Any wineries that want to get a WO appellation must submit their wines for regular chemical and sensory evaluations during each stage of the wine-making process.

SAMPLE OF SOME NEW WORLD WINE LABELS

CALIFORNIAN LABEL

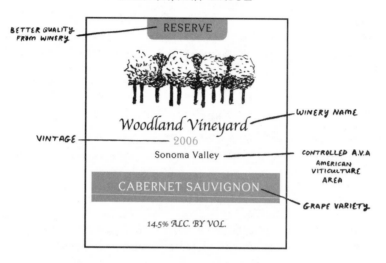

BETTER QUALITY FROM WINERY

RESERVE

Woodland Vineyard — WINERY NAME

VINTAGE — 2006

Sonoma Valley — CONTROLLED A.V.A AMERICAN VITICULTURE AREA

CABERNET SAUVIGNON

GRAPE VARIETY

14.5% ALC. BY VOL.

CANADIAN LABEL

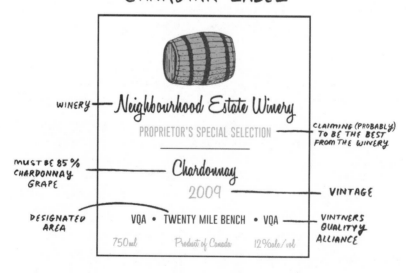

WINERY — Neighbourhood Estate Winery

PROPRIETOR'S SPECIAL SELECTION — CLAIMING (PROBABLY) TO BE THE BEST FROM THE WINERY

MUST BE 85% CHARDONNAY GRAPE — Chardonnay

2009 — VINTAGE

DESIGNATED AREA — VQA • TWENTY MILE BENCH • VQA — VINTNERS QUALITY ALLIANCE

750ml   Product of Canada   12%alc/vol

CONCLUSION TO SECTION THREE: THE APPROACH

The third key to understanding wine is knowing which approach the winemaker has used. The approach is defined by two decisions. The first is whether to blend grapes (make a *cuvée*) or make a varietal wine. I included the history of Bordeaux and Burgundy to help you understand some of the history of each approach and the reasons a winery might decide to take one approach or the other; I find knowing some background information helps me remember certain facts. Besides just building admiration for the wines from those two regions, each approach offers some sound advantages. The blended route is safer in terms of consistency and ensuring fuller wines. But the varietal approach has a simpler focus and a certain beauty to its emphasis on nurturing just one grape variety. Of the two, I opt for *cuvées* because I find them more complex and therefore more interesting. Yet some of my most vivid memories come from occasions where a varietal took center stage. I'll leave you to come to your own conclusions.

The other main aspect of the approach to winemaking involves the pursuit of quality. Many wineries simply concentrate on making value-based wines. Thank God! If it were not for them, I'd be drinking a lot less wine (my doctor might be pleased). Some of these wineries actually make very good wines and are the most profitable businesses. Every time you find a real bargain (e.g., you paid $10 but it's comparable to a $25 bottle), it's a pleasure.

My real admiration goes out to the makers of quality wine. The amount of dedication that's required to engage in quality wine production is tremendous. The people who pursue this approach are usually fastidious and knowledgeable. I can just sit for hours listening to their explanations for doing this or that in the vineyard or winery. Being in the company of people obsessed with high quality is kind of mesmerizing.

To understand the logic in labeling, you need to be a bit of a student of geography. Every country has developed (or is starting to develop) regional boundaries to define prime grape-growing areas. Remember, in Europe, with the exception of Alsace and Germany, knowing the name of the region is your only clue as to which grape varieties are inside the bottle. This knowledge takes a little memory work, but it comes easily after a few bottles.

In the New World countries, the names of the grapes are almost always on the front or back label. The big difference from Europe is that the grape-growing and winemaking regulations are less rigorously defined. As a result, you need to depend more heavily on the producer's reputation rather than knowing the appellation of origin. This fundamental difference between Old and New World labeling is slowly disappearing. One day, all top wine-producing countries will regulate the main parts of the grape-growing and winemaking processes. The best wine laws will not be too rigid and will allow for some experimentation and gradual evolution of these controlled practices as knowledge grows.

CONCLUSION TO
WineSense THE BOOK

"Wine-drinking is no occult art to be practiced only by the gifted few. Indeed, it is not an art at all. It is, or should be, the sober habit of every normal man or woman burdened with normal responsibilities and with a normal desire to keep their problems in perspective and themselves in good health." — Allan Sichel, A Guide to Good Wine, 1952

I. The Three Keys

I love the above quote from Sichel (a very knowledgeable wine writer). It encapsulates one of the main reasons I wanted to write this book in the first place. My goal was to remove any of the snobbishness associated with wine and reduce the intimidation factor that many people experience when trying to approach the subject. I wanted to make enjoying wine an approachable and accessible endeavor for so-called "normal" people. I wanted to ultimately *unlock* the mysteries of wine.

The Basics: Section One was essential for getting started with the exploration of this fascinating beverage. One needs to understand the basics of making wine and learn the different styles that are most commonly produced. A little bit of history brings the whole field of study to life. As I mentioned previously, just knowing the historical background of a subject seems to make understanding it — and remembering the central facts — a lot easier. Finally, it was important to get a quick introduction into the language of wine tasting and some helpful hints on appreciating the wine itself. Tasting with some focused attention is the *best* route to truly understanding wine.

The Grapes: Section Two was simply a close look at the fundamental elements in the winemaking process. I liken the study of the wine grapes to the scientist's exploration of the basic particles of matter: Grapes are the "atoms" of wine. And once more I apologize for all the memorization, but it's helpful to commit the classic grapes to memory. After all, the eight classic varieties comprise over 75% of fine wine. Knowing their names and some aspects of their traditional characters — and some history! — is well worth the effort.

The Approach: Section Three of the book completes the loop for understanding wine. After learning how it's made and knowing its component parts, you only need to be aware of what decisions a winemaker makes and why. Is the winemaker using one grape or blending a few? Is he or she going for quality or trying to make a value wine? Knowing the winemaker's approach to wine production will help you understand his or her goals and give you a hint as to what to expect in the wine (and the price). Remember, many of the clues to their approach will be found on the wine labels.

Going through these Three Keys will bring you to a firm grounding in appreciating wine. Walking into a wine shop will be an actual pleasure, somewhat similar to going into a great library. Reading the hundreds of wine bottle labels that are full of information, each telling a story about its contents, is a fine way to spend some time.

After having read this book on wine, I hope you're well on your way to a life-long journey of discovery. Explore the gifts wine can offer. I encourage you to frequently drink wine with family and friends. Celebrate. Introduce people to your favourites and tell them about all of its attributes. Share your newfound knowledge . . . your WineSense.

GLOSSARY

Acidity: All wines contain acids that come from the grapes (mainly tartaric and malic acids). Acid is an essential component of a wine, as it gives the wine structure and a refreshing zip. Insufficient acidity results in a flabby, uninteresting wine.

Appellation: This term refers to the *place* where the grapes were grown. It is always a controlled geographical name.

Aroma: Generally used to describe the smell of the wine *as a result of the grapes.* It is different from the *bouquet,* which is the smells that come from bottle-aging. However, many wine books use the terms interchangeably.

Auslese: The German label term for *Qualitätswein mit Prädikat* (QmP) wines made from carefully selected late-harvested grapes. These wines are sweet.

Barrel-aging: Many wines, mostly red, are aged in wooden barrels after fermentation. Traditionally, these barrels hold approximately 60 gallons and are made from oak; French and American are the two types of oak most commonly used. They are called *barriques* in Bordeaux.

Barrel fermentation: Some wineries ferment their grape juice in wood. This is typically done with mostly red wines and with only some white grapes, notably the Chardonnay. Barrel fermentation gives added complexity to the resulting wine.

Beerenauslese: German sweet *Qualitätswein mit Prädikat* (QmP) wines made from individually selected grapes that are often covered in the noble rot caused by *Botrytis.*

Blanc de blanc: White wine made from white grapes. A meaningless term in most cases, except for Champagne, where it means only the Chardonnay grape was used.

Blanc de noir: White wine made from red grapes. Certain styles of Champagne/sparking wines are made from only red grapes.

Botrytis cinerea: This fungus causes the noble rot (called Edelfäule in Germany) that attacks the grapes before harvest. It typically occurs in humid conditions and will dry out the grapes, thereby concentrating the sugars. The fungus is important for making Sauternes and other dessert wines. A wine that has been affected by this rot is said to have been *botrytized.*

Bouquet: Smells that are a result of bottle-aging. Generally not similar to grapey smells.

Canes: The shoots that come from the main branches of the grape vine. Some are removed, while others are trained onto trellises.

Canopy: The upper part of the grape vine, especially the leaves. The canopy must be carefully managed, starting with the trellis design and including pruning and thinning of the vine.

Cava: Sparkling wines from Spain made in the traditional Champagne method.

Chaptalization: The addition of sugar to the grape juice prior to fermentation. This practice is legal in many European countries, but it is carefully regulated. It is not allowed in California and many other New World regions. The ultimate purpose of chaptalization is to boost the alcohol level of a wine. It also adds body to the wine.

Charmat process: A bulk method for making sparkling wines. It's less complicated than the Champagne technique and generally results in poorer-quality bubbly.

Chef du cave: Cellar master or winemaker in French.

Claret: A term used to describe red wines. It originated from medieval times and was used to describe lighter coloured reds from Bordeaux (French spelling: *clairet*).

Crémant: Sparkling wine from France (not including Champagne).

Clos: French word for a walled vineyard.

Cru: This word means "growth," but it's used to describe a small area or vineyard and usually denotes top quality (e.g., *premier cru*).

Curvée: Literally, "blend." The practice of using more than one grape variety to make a wine.

Decanting: The practice of emptying a bottle of wine into another container, known as a *decanter*. Decanting is usually done for unfiltered wines or older vintages. The correct method is to place the bottle vertically 24 hours before decanting; then carefully open the cork and pour the wine into the decanter without disturbing the wine, leaving any sediment at the bottom of the bottle.

Dessert wine: Sweet wine made from overripe grapes. Sometimes referred to as "Late Harvest." Includes ice wines.

Disgorging: Removing the yeast sediments that resulted from the second fermentation (in bottle) in the Champagne-making process.

Dosage: The practice of adding sweet wine or grape juice to sparkling wines prior to bottling.

Dry: When all the sugar has been converted to alcohol, the wine tastes *dry*. Most table wines are dry.

Earthy: Describes wine that smells of earth or rotting vegetables.

Eiswein: German term for *ice wine*, which is wine made from frozen grapes. Very sweet and complex. Delicious!

Enology: The study of winemaking; spelled *oenology* in Europe.

Estate-bottled: Describes a wine that had its grapes grown on the same property where the wine was made and bottled. Usually an indicator of a quality wine.

Fat: Describes a full-bodied wine. Sometimes lacking in acidity and general balance.

Fermentation: The result of yeast cells consuming sugar and producing alcohol and carbon dioxide as a byproduct.

Filtering: Most wines are filtered before bottling to remove tiny bits of skins or yeast cells left in the wine after fermentation. Some top wines are not filtered, and you can find a sediment in the bottle. Some people claim filtering removes some flavour.

Fining: The practice of clarifying wine by adding various substances (egg whites, gelatin, isinglass, or bentonite) that attract loose sediment and take it to the bottom of the storage tank.

Finish: The lingering aftertaste from wine. Quality wines tend to have *long* finishes.

Flowery: Describes wine that gives off a pleasant floral aroma/bouquet.

Fortified wine: A wine with added alcohol, resulting in a wine that has a total alcohol level of 16–20%. Adding alcohol will stabilize a wine, meaning it won't deteriorate quickly if exposed to air or during normal aging. Port, Sherry, and Madeira are the most common examples of fortified wine.

Full-bodied: Describes wine that has a big flavour, complexity, a "meaty" texture, and high viscosity. Full-bodied wines are often high in alcohol (12–15%).

Grafting: Attaching a shoot of one plant to the root of another (often done with fruit trees). It was the solution to the phylloxera pest.

Grand cru: A term that means "great growth." It is used in Burgundy, Alsace, Champagne, and Saint-Émilion to denote a higher-quality vineyard.

Grassy: Describes wine that smells of hay or freshly cut grass; a typical aroma found in wines made from the Sauvignon Blanc grape.

Green: Describes wine made from grapes that were not fully ripened. Often, these wines are quite acidic. In some cases, like Portugal's Vinho Verde, greenness is considered an attribute, as the wine has a nice, crisp, fresh taste with good fruit flavours.

Heavy: Describes wine that is high in alcohol or sugar that is not balanced by some acidity (sometimes described as "fat").

Hybrid: A vine that is a cross between two species. Typically, it is a cross of the European *vinifera* grape vine and one of the American native species. One example, Vidal, is used widely for ice wine in Canada.

Ice Wine: See **Eiswein.**

Jeroboam: A large bottle, often used in Champagne; equivalent to four 750-milliliter bottles.

Kabinett: Literally translates as "cabinet." A German QmP-quality dry white wine.

Keller: German term for "cellar."

Kir: A beverage where currant liqueur is added to a dry white wine. Delicious sipper.

Late harvest: Grapes that are harvested later in the fall, usually making sweeter wine because the grapes are riper and therefore higher in sugar. Sometimes affected by noble rot (botrytis) as well.

Lees: A result of fermentation, lees are mostly dead yeast cells. Wine is sometimes aged on the lees (*sur lie* in French) to get added complexity.

Legs: Tears or rivulets that run down the sides of a wine glass. A result of glycerin and alcohol. Might indicate a fuller-bodied wine but says nothing about actual quality.

Length: Pronounced lingering aftertastes or finish. "This wine has tremendous length" is a common expression used to describe great wines.

Magnum: A bottle size equivalent to 1.5 liters (two normal sized bottles).

Malolactic fermentation: A secondary, bacterial fermentation. After the initial fermentation (the conversion of sugar to alcohol), wine sometimes goes through a secondary fermentation as a result of beneficial bacteria that converts malic acid to lactic acid. This fermentation tends to soften a wine (making it less acidic) and is commonly allowed to occur with most reds and some whites like Chardonnay grape–based white wines.

Meritage: A term used to describe blends of the grapes that are used in Bordeaux. Meritage reds have combinations of Cabernet Sauvignon, Merlot, and Cabernet Franc (sometimes Malbec and Petit Verdot are included). For whites, the combination is Sauvignon Blanc and Sémillon (and sometimes Mascadelle).

Microclimate: There are varying interpretations of this term, but it basically refers to unique climates of small regions — even single vineyards — that are different from the surrounding area.

Mildew: Fungal disease that attacks the grape vine regularly. There are two main types of mildew: downy and powdery. They are the bane of grape growers.

Must: Grape juice resulting from the pressing process prior to fermentation.

Noble rot: The *Botrytis cinerea* fungus that attacks grapes and dries them out, turning them to virtual raisins.

Nose: A term used to describe the smell of wine. "A very odd nose" describes unexpected and/or faulty smells.

Oenology: The study of winemaking; also spelled *enology*.

Oxidized: Describes wine that was exposed to too much air (a result of poor winemaking techniques or faulty bottle corks). The wine smells slightly burnt or stale, often like a poorly made Madeira or Sherry.

Phylloxera: A disease that wiped out most of the vineyards in Europe and California in the mid to late 19th century. The culprit was a tiny louse that fed on the roots of grape vines. Luckily, North American root stock was resistant to the louse. Today, the majority of vineyards, worldwide, are planted with North American roots to which European vines *(vinifera)* have been grafted.

Pruning: Grape vines are prolific plants and need to be tamed. Each year, grape growers need to prune the plant, cutting back the previous summer's growth. A minimum number of branches (canes) are spared for the following season.

Racking: After fermentation, wine is stored in large vessels. Periodically it is *racked,* or transferred to another tank, leaving sediment behind (mostly the *lees,* which are dead yeast cells).

Raisin: An overripe grape. Some wineries dry out their grapes after harvest to concentrate sugars. This is done most famously for Italian Amarone wines. The practice is known as *appassimento.* Sometimes wines made from overripe grapes actually have a raisiny taste, which can be a fault.

Reserve: A term often used on wine labels. When used by reputable producers, this means better wine put aside from the bulk of their production. The wine should be better quality, but the term is no guarantee.

Riddling: During the second fermentation of wine in Champagne, which is done in the bottles, sediment is produced. In *riddling,* the bottles are shaken slightly and progressively tilted, neck down, so the sediment collects in the neck of the bottle; later, the sediment is frozen so that when the cap is removed, the frozen sediment pops out (this process is called *disgorging*). The remaining Champagne is then topped up and properly corked.

Round: Describes a balanced wine with the right amount of acids, fruit, tannins, and alcohol. A round wine is smooth tasting and has full flavours.

Solera: A series of stacked casks arranged in rows used for making Sherry. The wine is transferred along the rows, from top to bottom, as the Sherry from different years blends and ages.

Sommelier: A knowledgeable wine waiter. Sommeliers may have various certifications.

Spätlese: A Qualitätswein mit Prädikat (QmP) German sweet wine. The term means a late selection of grapes.

Sur lie: A practice of leaving wine on the lees or sediment for a short time prior to further aging or bottling.

Sulphur: Used for preventing oxidation and also as a preservative. Sulphur has been used since antiquity as a tool for controlling fungus in the vineyard. No wine is completely free of sulphur (or sulphites) because sulphur is a natural byproduct of fermentation.

Table wine: A dry wine (as opposed to a sparkling, dessert or fortified wine) that's usually 10–14% alcohol.

Tannin: A type of compound found in wine, mostly red, that comes from grape skins and oak barrels. The same compound is found in tea. It tastes a little bitter and can cause you to pucker (as with strong tea). Tannins act as a preservative, and as a wine ages, they soften and transform into more-complex compounds, which produce interesting smells known as the *bouquet* of the wine. It is the tannins that help make red wine an excellent accompaniment for red meats.

Terroir: A complicated term that refers not only to the soil but also all characteristics of the local climate, soil content, aspect of the vineyard, slope, and elevation.

Thin: Describes a watery tasting wine, low in fruit and often low in alcohol.

Thinning: The practice of removing leaves from the grape vine to allow the grapes to fully ripen. Part of canopy management. Can also refer to removing unripened grape bunches before the harvest.

Treading: The traditional method of using your feet to crush the grapes (known as *pigeage*). Still practiced in a few wineries, notably for Port, and often accompanied by a celebration. Bacchus would be pleased!

Trockenbeerenauslese: A type of sweet German wine. Part of the Qualitätswein mit Prädikat (QmP) category and made from dried, botrytis affected grapes. The wine is an amazing dessert all by itself. (Note: *Trocken* also refers to a *dry* German wine.)

Ullage: The space between the cork and the wine in a bottle. If it is more than a half inch, the wine may be spoiled (i.e., oxidized and/or turning to vinegar).

Unfiltered: Describes wine that has not gone through any filtering after fermentation. Many people say an unfiltered wine maintains more flavour. You will usually find more sediment at the bottom of the wine bottle, but it's harmless; decanting these wines is a good idea.

Varietal: A wine made from one grape variety.

Vegetal: Describes a wine that smells of vegetables, specifically bell peppers, asparagus, and/or sautéed mixed vegetables. Sometimes considered a fault.

Viniculture: The art and science of winemaking.

Vinifera: The European grape species *(Vitis vinifera)*. For making wine, these species are the highest quality grape vines in the world.

Vintage: Wine made from a single harvest or year. In Champagne and Port, some years are not good enough to produce a "Vintage" Champagne or Port; this wine is then used only in blends of various years.

Viscosity: A syrupy characteristic. Alcohol is viscous by its very nature.

Viticulture: The science and study of grape growing.

Volatile acidity: Primarily refers to acetic acid, which is a vinegar acid. It naturally occurs in wine, but if it grows to the point where it can be detected, the wine is considered flawed; it is "vinegary."

Woody: Wine aged in wood sometimes has a strong, woody aroma. It can be offensive, versus a pleasant hint of oak.

Yeast: A single-celled organism that feeds on sugar, resulting in a conversion of grape juice into alcohol and carbon dioxide: wine! Some wines, especially Champagne, smell like freshly baked bread and are described as "yeasty."

Yield: A measure of productivity in a vineyard. Normally, the harvest is stated in terms of tons per acre.

Appendix #1: WINE & FOOD

"Wine makes every meal an occasion, every table more elegant, every day more civilized." — André Simon, *The Commonsense of Wine*

As far as I'm concerned, the title of this appendix is a bit misleading. For me, wine *is* a food. It has many healthful properties and is integral to many meals. I've often described wine as more like a complimentary "sauce" in its role at the dinner table. Wine is also a marvelous ingredient in many recipes. So let's discuss the *other* foods it goes well with as an accompaniment.

First, I'll lay out a few rules of engagement to give you a general understanding of wine's role. I call these rules *Desautels' Axioms,* named after yours truly. They're designed to give you guidance in choosing the right wine for a meal . . . or, conversely, to help you create the right meal for your wine. After a while, these axioms will appear to be common sense as you gain experience matching wines and food.

Desautels' Axioms

I. HARMONIZE

Strategy: Try to match similar characteristics in each item.

Examples:

- Sweet with sweet: Dessert wine with dessert or chocolate
- Light weight with lighter food: Fruity white wines are excellent with light meat (e.g., chicken) or vegetarian dishes
- Strong wine with full-tasting foods: Full-bodied reds go well with red meats and game

II. CONTRAST

Strategy: Pair opposite attributes.

Examples:

- Acid versus cream: Use a refreshing white wine to cut through cream sauces
- Fruit opposed to spice: A juicy, full-bodied red with a spicy meat dish
- Sugar versus strong cheese: Dessert or sweet fortified wines are a perfect foil for blue and old cheeses

III. GEOGRAPHY

Strategy: Find regional wines to match regional foods (and use the same wine in the recipe!).

Examples:

- Wines from Burgundy for Beef Bourguignon
- Coastal white wines with seafood
- Italian reds with pasta dishes
- Serve the wine used in the recipe

IV. SUPPORTING ROLE

Strategy: Let only one item shine.

Examples:

- Serve a great wine like Château Lafite with a straightforward roast beef
- Drink competent wines with exemplary gourmet meals

Note: If one of the two, food or wine, is *extraordinary,* then just focus on that one so you can truly appreciate the experience. Don't allow a great wine to compete with some amazing food (and vice versa).

V. ICONIC COMBINATIONS

Strategy: Forget all the rules!

Examples:

- Intensely complex sweet wines with foie gras (rich pâté)
- Spicy, mineral wines (e.g., Gewurztraminer) with smoked salmon
- Vintage Port or Madeira with blue cheese like Stilton
- Champagne with everything, or nothing!

Classic Food Matches For Table Wines

WHITE	Light-Bodied Neutral	Salads, Asian, vegetarian, delicate fish, lighter shellfish, pork, goat cheese
	Fruity & Flowery	Vegetarian, Japanese, Chinese, light pasta, chicken, fruit, soft cheese, most fish and shellfish
	Complex Full-Bodied	Vegetarian, Spicy Asian, pasta, pizza, pork, game bird, all fish including salmon, shellfish, fruit, hard cheeses
RED	Light-Bodied (& Rosé, Fruity Reds)	Spicy Asian, Japanese, vegetarian, pasta, pizza, shellfish, pork, chicken, veal, rabbit, salmon, tuna, hard and soft cheese
	Mature, Complex &, Balanced	Pasta, pizza, stews, spicy dishes, beef, lamb, game bird, sausages, burgers, hard cheeses
	Concentrated, Powerful, & Full-Bodied	Pasta, light curry, beef, lamb, venison, caribou, southern BBQ, hard cheeses, old cheddar

Note: These matches are neither definitive nor exhaustive. Many cookbooks have wine matches for their recipes. Always experiment. Learn.

Appendix #2:
BOTTLES & GLASSES

Bottles: There are many types and shapes of bottles used for wine these days. I have picked seven basic/standard styles.

1. *Bordeaux:* This shape is traditionally used for Cabernet Sauvignon, Merlot, and most red wine blends. It is also used for many dry whites.

2. *Burgundy:* This bottle is typically used for Pinot Noir and Chardonnay wines. It is also used for Rhône style wines made from the Syrah/Shiraz grape.

3. *Fiasco:* A traditional bottle used in Italy. The bottom two-thirds of the bottle is wrapped in straw to protect it from breakage.

4. *German:* A typical bottle for Riesling style wines.

5. *Fortified:* The traditional shape used for Port and Madeira.

6. *Onion:* Used in Spain and Portugal for young wines, often rosés; a similar shape is used for reds in Southern Germany.

7. *Champagne:* This is the commonly used shape for sparkling wines. The glass is usually thicker than other wine bottles and has a small indentation in the bottom (the *punt*).

Note: The shape of the bottle (see Fig. 13) will often give you an indication of the style of the wine inside. For example, using the number one bottle above usually means the winemaker is striving to make a Bordeaux-style wine.

Fig. 13

Glasses: There are four main types of wine glasses, plus many variations in size and subtle variations in the shapes. Below is an example for most wines: 1) fortified, 2) red, 3) sparkling, and 4) white and rosé (see Fig. 14).

Fig. 14

Appendix #3: SOME TIPS FOR WINE STORAGE

Great wines will improve with age. The first aging is done at the winery in oak barrels (see Fig. 15). The oak allows slow oxidization, which will "soften" a wine, smooth out the rough edges in taste, and start the wine on its way to maturation. The second stage is completed in the bottle. Here are five basic rules for designing a storage area:

1. *Light:* Ensure the room is dark; light is an enemy of wine.

2. *Moisture:* A little humidity (approximately 70%) will stop corks from drying out .

3. *Temperature:* High or low temperatures are not good for wines. Keeping wine at a constant temperature is essential; approximately 50°F or 10°C is ideal.

4. *Movement:* Protect the wine from vibrations; ensure your wine area is away from machinery and certain appliances.

5. *Orientation of the bottles:* Lay the bottles horizontally on a rack so the wine stays in contact with the cork. Ideally, have the neck of the bottle a bit higher than the bottom.

Fig.15

Appendix #4: AROMA WHEEL

(University of California — Davis Campus)

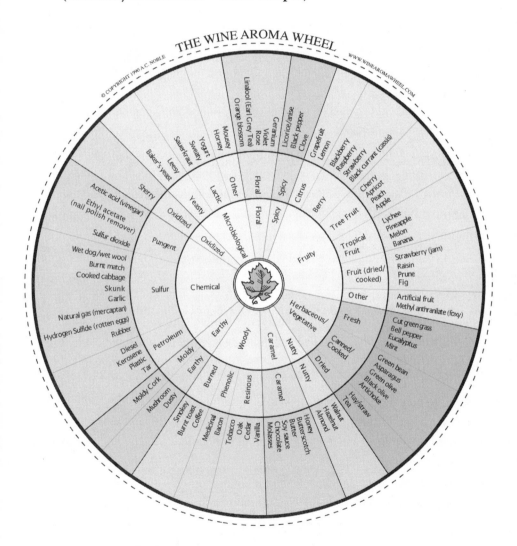

(Wine Aroma Wheel copyright 2002 A C Noble
http://www.winearomawheel.com/)

RECOMMENDED READING

For Pleasure

Vintage: The Story of Wine — by Hugh Johnson

Adventures on the Wine Route: A Wine Buyer's Tour of France — by Kermit Lynch

Red, White, and Drunk All Over: A Wine-Soaked Journey from Grape to Glass — by Natalie MacLean

Liquid Memory: Why Wine Matters — by Jonathan Nossiter

Vintage Pellegrini: The Collected Wisdom of an American Buongustaio — by Angelo Pellegrini

The Noble Grapes and the Great Wines of France — by André Simon

Resources

Hugh Johnson's Modern Encyclopedia of Wine — by Hugh Johnson

The World Atlas of Wine — by Hugh Johnson and Jancis Robinson

Alexis Lichine's New Encyclopedia of Wines & Spirits — by Alexis Lichine

The Wine Bible — by Karen MacNeil

Parker's Wine Buyer's Guide — by Robert Parker

The Oxford Companion to Wine — edited by Jancis Robinson

Vines, Grapes and Wines — by Jancis Robinson

The Sotheby's Wine Encyclopedia — by Tom Stevenson

Larousse Wine: The World's Greatest Vines, Estates, and Regions — by multiple authors

About the Author

Bob Desautels describes himself as a wine enthusiast who is "more gourmand than gourmet in foods and wines." Bob's formal education includes a Bachelor of Commerce and Master of Arts in Philosophy. At the University of Guelph in Ontario, he taught for ten years in the School of Hotel & Food Administration's Bachelor of Commerce program. Although he taught eight different courses at the university, his favourite was the wine course. One year he took a group of students to Europe to learn about the vineyards and wineries of Germany and France. Bob often gives talks on wine and has taken numerous trips to Ontario's wine region with small and large groups.

Bob is also a successful restaurateur. He opened one of Canada's first wine bars in 1985. Recently, the Ontario Hostelry Institute named him Restaurateur of the Year. As president of a hospitality business, The Neighbourhood Group of Companies, he oversees three restaurants along with the Group's signature "Wines from the 'Hood" and "Taste of Ontario" food products. Bob has been a champion of locally crafted foods and beverages since 1990. He is also a leader in sustainable business practices; his company won a national Green Leadership Award in 2013.

.

Printed in Canada